REVEALED! HOW YOU TO STEAL YOUR OWN INFORMATION

WHAT YOU NEED TO KNOW TO PROTECT YOUR BUSINESS FROM SOCIAL ENGINEERING & CYBER ATTACKS

A plain-English, no-nonsense business owner's guide to the biggest cybersecurity threats out there today, and how to protect your business from risk.

Read this book and you'll discover:

- ✓ The biggest cybersecurity risks affecting businesses right now and how to protect your company from danger.

- ✓ The most common threats that cyber-criminals are currently using to break into your accounts, hijack your pages, and compromise your employees and the entire company as a result.

- ✓ Case studies that delve into the anatomy of the most common types of attacks to deepen your understanding, so you can identify threats before they do damage to you and your company.

- ✓ How cyberattacks have evolved with changing technology to get around new safeguards almost as fast as we invent them.

- ✓ Appropriate responses to cyberattacks versus how improper procedure can do further damage to your business.

- ✓ Future cybersecurity threats on the rise and what prevention measures you can take now to protect your business from compromise moving forward.

- ✓ What to look for in a cybersecurity vendor to guarantee you're getting the best possible defense against attacks.

- ✓ The latest software and hardware protections aimed at combatting current and growing threats targeting businesses like yours today.

Cybersecurity Expert & Master Technology Strategist:
Al Alper

author of the bestselling
Revealed! series of books

Plain-English, No-Nonsense Business Owner's Guides Written to Protect Yourself and Your Business from the Dark Underworld of the Internet

Al Alper
Absolute Logic
88 Danbury Road, Suite #1D
Wilton, CT, 06897
203-936-6680 • 855-978-7252
www.absolutelogic.com

Dedicated to Helping Prevent Business Owners From Becoming Victims of Cyber Crimes FOREVER!

New York * Connecticut * New Jersey * Massachusetts

All rights reserved. No part of this publication may be reproduced or transmitted in any form by any means, electronic or mechanical, including photography, recording or information retrieval system, without written permission from the author.
Printed in the USA. Copyright © 2016 Al Alper
ISBN

CONTENTS

Introduction 6
   The Evolving Threat of Social Engineering
   Attacks   6
   Protecting Your BUSINESS Today   7
   Social Engineering and Reverse Social
   Engineering Attacks   8
   What This Book Will Teach You   9

Chapter ONE 12
   An Introduction to Phishing   12
   Case Study: Google and Facebook   16
   Case Study: FACC   18

Chapter TWO 23
   An Introduction to Quid Pro Quo   23
   Case Study: Impersonating SSA   25
   How to Handle a Quid Pro Quo Attack   27

Chapter THREE 29
   An Introduction to Baiting Attacks   30
   Case Study: Stuxnet Computer Worm   32
   Case Study: GoDaddy Holiday Bonus   34

Chapter FOUR 38
   An Introduction to Pretexting   38
   Case Study: Business Email Compromise   39

Chapter FIVE 46
   An Introduction to Man-in-the-Middle Attacks   46
   Case Study: Alexa and Google Home   48
   Case Study: Lenovo and Superfish   53

Chapter SIX 57
   An Introduction to Cloud Vulnerability   57
   Case Study: Uber   59

Chapter SEVEN 64
   An Introduction to Tailgating   64

Case Study: Colin Greenless 65
Chapter EIGHT **70**
   An Introduction to Water-holing 70
   Case Study: Forbes.com 71
   Case Study: Department of Labor 73
Chapter NINE **76**
   An Introduction to Spoofing 76
   Anatomy of an IP Spoofing Attack 77
   Case Study: Dyn, Inc. 78
   Anatomy of an ARP Spoofing Attack 81
   Case Study: Metasploit Project 81
   Anatomy of a DNS Spoofing Attack 83
   Case Study: Malaysia Airlines 84
Chapter TEN **88**
   An Introduction to Smart Contract Hacking 88
   Case Study: Parity 90
   Case Study: Block.one 91
Chapter ELEVEN **95**
   An Introduction to Rogue Scanner Hacking 95
   Case Study: DarkAngle 97
   The Debate Over Rogue Scanners' Threat Level 99
Chapter TWELVE **103**
   An Introduction to AI Enhanced Threats 103
   Case Study: TaskRabbit 105
   Case Study: Instagram 107
Chapter THIRTEEN **111**
   How Cybersecurity is Changing to Help 111
Chapter FOURTEEN **117**
   Technical Terms Explained in Plain English 117
An Invitation to the Reader **120**

Introduction

The Evolving Threat of Social Engineering Attacks

The more entwined technology gets in our everyday lives, the more familiar even the least technologically-savvy person you know gets with common cyberattacks. That Nigerian Prince has been effectively outed as a worldwide hoax, and far less people are going to mindlessly give out their social security number to the first robocall that asks. Even ten years or so ago, this wasn't the case at all.

As people got more savvy at recognizing simple scams like those, cybercriminals had to get smarter in retaliation. Electronic inventions have gone through so many iterations in the past few decades alone that it would have been impossible to predict, thirty years ago, what problems face modern businesses in 2021. Looking ahead, we can't imagine the innovative ways cybercrime will evolve to get past the new and improved smartphones that will come equipped with features we can't even imagine yet.

The upside is that the Internet of Things is constantly improving to give us a faster, better and more interconnected experience with each other. The downside is that cybercrime is always changing too, thus top-of-the-line safeguards have a very short shelf life before they're obsolete, being dethroned for something better as cybersecurity clamors to stop the new threats—and around and around it goes.

All of this is just a way of saying that there's always a new Nigerian prince just around the corner; a new wolf in sheep's clothing that target businesses like yours, convincing YOU to let them in!

What are you doing to make sure the door's locked tight so that they can't get in?

Protecting Your Business Today

Just because each threat won't last forever doesn't mean it's futile to take precautions that will save your business now. You may be asking:

- Why is it useful to take short-term precautions in the absence of a long-term solution?
- What should I do if I notice a virus or malware has already gotten through my network?
- How can I tell the difference between a potential cyberthreat and a genuine client, customer or vendor?
- How could any of this possibly happen to me?
- Is it too late to protect my business?

That's what this book will teach you.

Together we will explore the dangers of modern social engineering attacks by breaking down the anatomy of **twelve** major cyberattacks hurting businesses just like yours today. Using case studies, you'll learn to identify the early signs of these cyberthreats to reduce the risk to your business.

Then this book will teach you how to react to potentially dangerous situations so in the future, if one of these attacks happen to your business, you can save yourself and your cybersecurity team from a bigger problem by minimizing the damage from the get-go.

Social Engineering and Reverse Social Engineering Attacks

Before we dive into different examples of social engineering attacks, let's first define what it is we're talking about.

In most cases, cybersecurity threats result from somebody in your company or connected to your network unwittingly giving out the information to compromise your accounts to an untrustworthy outside source, who then uses it to break into your system or otherwise steal, replicate or do damage to your data and information. This is called a **social engineering attack**.

Let's go back to our old friend, the prince.

If he sends one of your sales reps an email saying that the company has just won $10,000.00 and all she needs to do is fill out your bank information so he can wire the award, that's a social engineering attack.

But what if the prince isn't the first one to reach out?

Let's say that same sales rep is on a dating website and she gets catfished[1] by the Nigerian prince. She finds his profile, is intrigued by his riches and makes the first move. This builds their trust even deeper because she's the one who established contact, so when he asks her for the company bank information so he can invest, she's more likely to believe it. This is called a **reverse social engineering attack**.

The difference lies in who makes first contact. And when that first contact is you or your team, the likelihood that the attack will be successful is almost 100%!

---

[1] Catfishing refers to someone creating a fake profile online with the intent to befriend or date another person.

It's not enough to put up firewalls and add next-generation security at every point of entry into your network; you also need to train everyone in the company, including YOU to recognize what threats look and act like; the traps that are going to be a lot better concealed than our prince's email or a robocaller.

Then and only then will the game change, and instead of being just another statistic, another victim of a cyber-attack, you and your company will be a success story, a model for others to follow.

What This Book Will Teach You

- ✓ The definitions of twelve different cyberattacks most likely to threaten businesses like yours.

- ✓ Case studies of each threat to walk you through the initial point of entry, what happened in each instance and how they ultimately solved the problem.

- ✓ The appropriate response to each of these cybersecurity threats, what not to do, and why the wrong move can spell even greater disaster for your business.

- ✓ How social engineering attacks can affect any size organization, from family-owned businesses to massive tech conglomerates that you'd expect to be better prepared.

- ✓ What all of this means in relation to your small business.

- ✓ Reorient your ideas about Internet safety and the cost-benefits of advancing tech.

- ✓ How AI is beginning to play a role in cybersecurity and cyberattacks alike, and how this is projected to evolve in the near future.
- ✓ What the future of cybersecurity will look like in reaction and relation to these threats.
- ✓ Simple, everyday steps you can take to protect your business from giving out too much information.
- ✓ Change your approach to social engineering security on a long-term scale.

As technology continues to evolve and cybersecurity threats come, go and change, stay alert so you meet all future cyberattacks with a smarter defense.

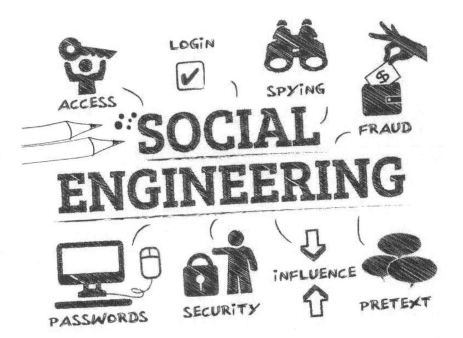

This Photo by Unknown Author is licensed under CC BY-ND

Chapter ONE

An Introduction to Phishing

Phishing has been an extremely popular tactic for cybercriminals since its introduction into the world in the mid-1990s. It refers to a type of hacking where the criminal steals private information by tricking you or somebody close to you, or with access to information in your business into giving it up directly or by giving up the credentials that gives them access to the information, usually through carefully crafted emails designed to make you trust the sender and willingly give them what they are asking for.

Not surprisingly, we're back to our Nigerian prince example again. Click on a spam email asking you to chat, and the next thing you know there's malware or a virus infecting every corner of your network.

As you know, these emails have gotten much cleverer and the other tricks are sneakier than a robocall asking for your social security number without any pretext—although even these have gotten sneakier, moving away from, "You won a prize!" to pretending to warn the victim about car payments or identity theft, so they can ask them to confirm their credit card or social security number. Still, most people recognize these scams for what they are.

How, then, did three-fourths of all businesses experience a phishing attack in 2020 (*Source: Tessian, 2021*)? A lot of criminals have realized the innate potential of global terror; capitalizing on fear surrounding COVID-19 has led some to fall victim to phishing when they might otherwise have been clearheaded enough not to make that kind of mistake. It just goes to show

how cunning—and manipulative—cybercriminals can be, and how easy it is to fall into their traps if they push the right buttons.

That's why it's necessary for businesses to stay informed, not just about the current forms of phishing threatening their business, but also how phishing scams have evolved over time so that they can better recognize up-and-coming threats as they arise in the years to come.

Here are some basic statistics to get you started:

- 91% of all cyber-attacks begin with a phishing email to an unexpected victim, as reported in Deloitte's 2020 Cyber Risk Advisory Report.
- According to a report by Kratikal, 85% of all organizations have been targeted by phishing attempts.
- Despite how common phishing attacks are, only 3% of employees report phishing emails to their higher-ups (*Source: Security Boulevard*), we like to call that the 3% club, demonstrating how crucial it is to equip your staff with the knowledge and protocols to handle these attempts to break-in to your network.
- Most phishing emails are sent by cybercriminals pretending to be from Apple (*Source: Check Point Research*). Scammers often use household names or department handles to increase their believability.
- CyberGuard360, a cybersecurity company specializing cyber awareness and training, found that cybersecurity training is more critical than many business owners think: Almost 78% of people who went untrained failed an experimental phishing assessment in 2020.

Clearly, it's important for your business to stay informed about up-and-coming phishing scams. Between cybercriminals getting smarter and using advanced techniques to infiltrate businesses, to a lack of reporting, to untrained employees posing a significant risk just by virtue of being ill-equipped to deal with unusual emails or calls, Businesses need to know everything they can about how phishing attacks work so they can protect their business from all angles.

ANATOMY OF A PHISHING ATTACK

While phishing techniques evolve at lightning speed, and tomorrows phishing attacks won't necessarily look the way they do right now, that doesn't mean it is any less important to learn how they work so you're protected. Impermanence does not mean unimportant.

Phishing attacks usually come by email; they are primarily focused on stealing log-in information, staff credentials or personal data so they can infiltrate your company. Rather than straightforward hacking like brute force attacks or sneaking in through a backdoor in the network, hackers ply their trade with phishing attacks that trick the person into giving up the information willingly, without any effort whatsoever.

Hackers most often take on the persona of authority figures like managers, supervisors or people they believe need it like the technical support department, or even outside parties like the bank, finance company or a vendor they know the company does business with; some trusted (*supposedly*) source that people are predisposed to believe should have or need to know the information they are asking for.

A great rule of thumb is don't give away your social security number or credit card or login information or really any information to just anyone; when in doubt, don't give it away at all!

Still, this doesn't mean that your employees are safe as long as they're not replying to emails. Phishing also happens over the phone, text or online—everyone must be on their guard whenever they're talking to someone if they don't know for sure that they're a trusted source.

Criminals are getting smarter. Instead of simply sending out generic emails looking to trick unsuspecting victims into giving out the keys to the kingdom, or vague threats under the name of the Internal Revenue Service or a large bank to a massive list of potential victims, they target specific businesses or people.

This very popular form of phishing attack is called **spear-phishing** and poses significant dangers since it has proven very effective.

It is much more effective than a mass, vague threat or request for information of a traditional phishing attack. Spear-phishing is the targeting of specific people using information about them or their families and friends that they, themselves have given out, typically on social media or might have been exposed in another breach, like that of Experian or Equifax and found on the Dark Web. This information is then used to craft language that implied they know them personally.

By doing research on someone beforehand, cybercriminals can launch a personalized attack on a vulnerable employee designed specifically to trap them. Usually, the message seems to be from someone they know, because the criminal has detailed information about them, their job, their friends and/or their family.

A great way to prevent a successful spear-phishing attack is by showing employees how to be careful about what information they share on and off the clock, so they don't fall victim to spear-phishing attacks that affect them and the entire company.

Limiting or restricting access to employees to check their social media or less-secure personal accounts on company time can help reduce the risk that they may give out information that can be used to launch a successful spear-phishing attack that may put your company at risk and expose you to a major liability.

Learn more about the importance of protecting social media from cybercriminals in the third installment in our series: *REVEALED! The Cyber Threat of Social Media.*

Case Study: Google and Facebook
WHAT WENT WRONG?

Even the biggest companies with unlimited resources at their disposal are at risk of falling victim to successful phishing attacks.

Google and Facebook are massive technology conglomerates employing some of the best and brightest people in the world who have vast knowledge and experience in cybersecurity, and still they both fell victim to a fairly basic phishing scheme that's no less cunning just because it's so straightforward.

In a scam that ran from 2013 until 2015, Evaldas Rimasauskas, and a team working with him, defrauded more than $120M from the companies together. They set up false business accounts and used a series of fake invoices, contracts and letters to pull off the attack and defraud these companies.

By pretending to be from Quanta Computer, a real company based in Taiwan who has done plenty of business with both Facebook and Google in the past, Rimasauskas sent invoices to the employees who usually worked with Quanta in the past. Rather than asking questions, the workers paid up.

Ultimately, this mistake cost the companies over $100M over the course of those three years. Rimasauskas forged Google and Facebook executives' signatures on the invoices, Quanta employees' signatures, and even went so far as to stamp the documents with fraudulent seals to trick the banks into complacency. For three years, no one asked questions as Rimasauskas transferred money to accounts all over the world and he was free from the SEC.

Eventually, Google detected fraud on their accounts and reported those discrepancies to the police. They recovered all of the lost funds, and Facebook recouped the majority of their stolen money as well. Rimasauskas pled guilty to wire fraud and was sentenced in 2019. He's currently serving five years.

WHAT SHOULD HAVE HAPPENED?

Remember the 3% club, how only 3% of phishing attacks get reported to management? Now you begin to understand how Google and Facebook both suffered by not giving proper and regular security awareness training to their employees.

It's especially important for those who handle company finances to be well aware of the potential dangers, and for business owners to train them to ask questions, be wary of any unexpected transactions, and that there's no harm in double-checking with higher-ups before paying out massive sums to an unknown source—even if they look trustworthy. Just because you've given hundreds, thousands or millions of dollars to a company in the past doesn't mean you should blindly hand them money whenever they ask, without first confirming what services were rendered and that they got the job done right.

In short, be wary of emails asking for any money without first confirming why they need it, that they earned it and that the account holder knows what the payment is for. Set up an internal system to double-check that goods or services were properly rendered and that the final amount is accurate before making any payments, particularly in volumes that high.

Train employees to be wary of monetary transactions, not to give up private information, and to ask questions when they spot a red flag. Make sure your company is part of that 3% club!

Case Study: FACC
WHAT WENT WRONG?

In the above example, Google and Facebook were targeted as entire companies, but sometimes employees are singled out on an individual basis. Take what happened to the Austrian company, FACC.

In 2016 FACC, an aerospace manufacturing company with major industrial clients like Airbus, Boeing and other big names reported they'd been involved in a spear-phishing scam that cost the company 42M euros—that's $47M in American money, and only 10.9M euros were blocked from transfer!

FACC were victims of what is commonly known as a **"fake president"** scheme.

Essentially, cybercriminals used domain spoofing (see the *Glossary*) and targeted a specific employee, asking them to transfer money immediately to cover an acquisition the company was making. The email *appeared* to come from the CEO.

By making it seem urgent and from someone important in the company, workers are less likely to pay attention to small details that would signify a spear-phishing attack, such as minor discrepancies in the domain name or email signature.

The employee transferred the money to the fraudulent bank account under the impression that they were acting at the request of the CEO, Walter Stephan.

The CFO, Minfen Gu, was fired in February following the attack. After a long board meeting in May, Stephan was also let go by the board of directors because he "severely violated his duties," according to an official statement. Although the company didn't disclose what his involvement in the case was, a lawsuit filed in 2018 against both the former CEO and CFO accused them of taking improper actions to prevent something like this from happening in the first place. The lawsuit was ultimately thrown out, but it does demonstrate the disastrous consequences that can happen both personally and professionally following a successful spear-phishing attack.

WHAT SHOULD HAVE HAPPENED?

There is an old saying that the best defense is a good offense. When fighting cyber-criminals, the best defense is preparedness. Although the amount that FACC was defrauded *was* unusually high for this type of attack—it's not unheard of.

Just one year before the attack against FACC, the FBI gathered that U.S. companies lost $246M to fake president scams; and remember that that only covers what was reported. Plenty of companies don't want to admit that they were a victim of cybercrime since it can affect their reputation and cause them to lose customers and investors; experts believe that the actual statistics are significantly higher.

When it comes to spear-phishing attacks, criminals can be even harder to detect than usual because they have a specific target in mind, enough background on that individual to know who they are most likely to let their guard down around—for example, the CEO of your company.

Although they weren't ultimately found liable in court, the consequences that did occur for Stephan and Gu are enough to demonstrate the importance of setting up strong internal systems so that nothing like this happens to you or your employees. Attention to detail is everything: The ability to spot a period or an underscore that's not right in a fraudulent domain name doesn't seem like a difficult skill to learn, but all it takes is one mistake to lose thousands, if not millions of dollars, and potentially put you out of business.

So how can you protect your company from the fake president attack?

The answer can be summed up in the famous quote by former President Ronald Reagan, "trust but verify."

Establishing a protocol where any request for money, especially large ones must be confirmed by a phone call to or from the person making the request, or require that a second senior person verify the amount and transaction are correct.

Training is an essential element here as well.

Making certain that all employees know the company's policies and procedures, and regularly testing them to ensure they know them is an essential first step in building a defensive line to protect the company.

Include a robust security awareness training program that teaches employees to spot domains that have been spoofed, and reinforces it with regular "reminders". Platforms like PIISecured, available from your IT service provider can automate all of this and provide you the visibility you need to know who is paying attention and who is most likely to put your company at risk.

Developing a proper training and awareness program throughout your organization, and monitoring who is attentive can be the difference between the next prevention success story and a $50M mistake.

Chapter TWO

An Introduction to Quid Pro Quo

In Latin, quid pro quo translates directly to "something for something." That's exactly what a quid pro quo attack is: A cybercriminal holds your information hostage in some way, and they promise to give it back if you pay them off. Sometimes they hold it hostage by stealing and storing it and then deleting it from your systems, other times they use ransomware to hold it hostage.

If you know anything about ransomware, you can already see how these often work together. Ransomware is a type of malware that encrypts your information, making unreadable without an decryption key, effectively blocking user access to their information until they pay a ransom, whereas quid pro quo attacks encompass a range of tactics that don't necessarily have to involve ransomware or other malware being installed.

Most commonly, criminals pretend to be part of the Social Security Administration, or SSA, and say that they have an issue on their end so they need the unsuspecting victim to confirm their social security number with them. This happens a lot on an individual level, but businesses are more often targeted by someone pretending to be with the I.T. department. They call staff members about "necessary" upgrades or installations, or a quick fix for computer problems that the employee hasn't even noticed yet. Since most peoples I.T. knowledge is limited, the employees often give up access to their computers, which gives the criminal a chance to install malware right then and there, in front of them.

It's scomplicated, but it doesn't have to be. When you run a business, you're the one who's most invested in your company. Hourly, line or lower-level workers are much more susceptible to these tactics because they aren't as invested in the company's success and security as you are, and some may simply just not care as much as you do about protecting the company. They are more likely to make careless mistakes if they don't have the same passion, proper incentive or training to protect company information to the absolute best of their abilities. It's your job as their leader to guarantee they're ready to handle it when an attack is launched against your company and know what to be on the look out for, and what to do.

ANATOMY OF A QUID PRO QUO ATTACK

Often quid pro quo attacks take the form of someone impersonating I.T. reaching out to an unsuspecting victim inside the company. The cybercriminals will spam as many direct numbers in a company as they can find and offer the exact same I.T. assistance to everyone who picks up. This serves two purposes. For one, employees are less likely to get suspicious of the calls if they know their coworkers all received similar messages. For another, it increases the likelihood that they will find someone who says yes and lets them into the system. They only need one.

Once they've tricked a staff member into believing they are from I.T., the attacker generally tells the victim that they have a software update, installation or fix to a problem that the employee didn't know they needed. The employee lets them into the system, whether remotely or in-person, where they can then install malware or steal data. Just like that, the company's been compromised. The attackers make off with the valuable data they need to hold hostage until they get what they want.

Let's consider how quid pro quo attacks have played out in real life so you can better understand how to identify and avoid them.

Case Study: Impersonating SSA
WHAT WENT WRONG?

There has been an influx of reports coming in about people using the coronavirus as a pretext for stealing personal information. They're capitalizing off of widespread fear and confusion. The Federal Trade Commission recently disclosed that there is a particular scam going around wherein cybercriminals pretend to be from the Small Business Administration Office of Disaster Assistance.

The FTC warns businesses that they're being specifically targeted by purveyors of this scam, who send emails out stating that your business is eligible for a disaster relief loan of as much as $250K. Before you can get those necessary funds, though, they'll need you to confirm your date of birth and social security number.

This is just the tip of the iceberg. The COVID-19 virus has wrought terror, misinformation, and fear amongst everyone from your clients to your top-level employees. When you're in a financial bind caused by more than a year of a tragic pandemic, unsure what the future holds and constantly trying to run a business on unstable grounds, you're more likely to fall victim to scams that you might otherwise be clearheaded enough to spot for what they are.

It's not just fake representatives from the Small Business Administration. In addition to the SBA, scammers pretend to be from the IRS and Social Security Administration to try and get your bank account and identifying information.

Cybercriminals are very good; they're not just exploiting mass fear. They also use spoofing to fake their area code to seem legitimate; you're less likely to believe the SSA is calling from Oklahoma or an out-of-country number than if they had a D.C. area code. The SSA recommends calling their office at 1-800-772-1213 to confirm if you've genuinely been contacted by them or if it's just a scam.

Don't let fear dictate how you act and react. Take a deep breath and don't leave your common sense on the other side of the phone.

WHAT SHOULD HAVE HAPPENED?

Your social security number is absolutely necessary for criminals trying to steal your identity. Scammers will call insisting they need to confirm your identity, or they might set up fake websites where they ask you to enter your social security number.

You should never give your private information to anyone or anywhere that you don't completely trust. Ssa.gov is the only place where you should apply for a new social security card, and you should never give your social out over the phone. Due to spoofing technology, you can never really trust who's on the other line if *they* call *you*.

Quid pro quo scammers use false domains and official-sounding names to convince you that it's a legitimate contact, but always be suspicious. Don't give your bank account information, credit card numbers, social security or any other personal information to someone who you are not completely sure they are who they say they are, even if they appear to be coming right out of the capital.

How to Handle a Quid Pro Quo Attack

Common sense, training and a strong defense can help your business avoid quid pro quo attacks that come its way in the future.

For starters, don't give all of your employees equal, top security access, especially if you have several tiers within the organization. Only tell a select, trusted few important information that could compromise your business, like partners or the highest-level employees. The more people who know how to get into or disable your security, the more who can give remote access to cybercriminals trying to dupe the company out of money or confidential information. Your odds of someone making a mistake increases with each person you tell.

Next-generation antimalware software can save the day when you least expect it, but need it most. Installing appropriate safeguards on all of your machines will catch some of the malware that they install when they crack through, thus mitigating some of the damage in the event of a breach.

Quid pro quo attacks generally involve spamming as many members of an operation as the criminals can get a hold of. That's why training your workers is so important. Holding seminars about basic cybersecurity, putting them to the test with online assessments and getting in touch with experts in I.T. security will equip all of your workers with the tools they need to identify quid pro quo attacks and stop then them in their tracks. When anyone can be a leak in the business, everyone has to be prepared.

Most importantly, rely on your common sense. Quid pro quo attacks don't always come from someone masked as I.T. or the SSA; it can be winning a contest that you never entered or deals that you don't qualify for. If you get a call with an offer that seems too good to be true, do some digging before handing over your bank information. And when it comes to I.T. support, remember that people aren't always who they say they are over the phone and online. If you never spoke to them before, contact the person in I.T. you have spoken to before or your supervisor.

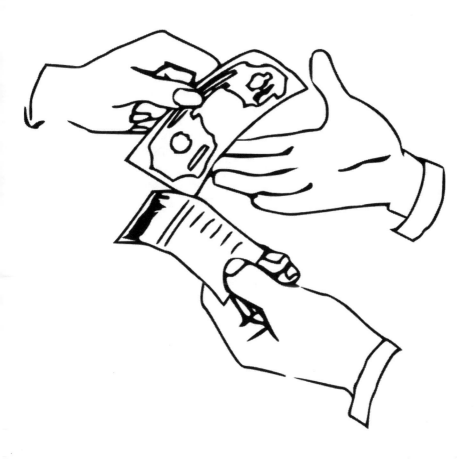

Chapter THREE

An Introduction to Baiting Attacks

Cybercriminals don't rely solely on threats to get targets to do what they want. Sometimes people can be deceived by the promise of something alluring, something that makes them curious enough to be susceptible to accepting malware-infected documents and other files.

Baiting attacks are traps designed to lure people into doing something they might otherwise be cautious about doing. They are a ruse where they promise a tempting prize or make claims designed to pique the victim's interest. They make terrific offers and craft language to convince people to open the infected link or malicious file.

The links or files can be spread through social media, in a hidden folder somewhere for them to find and open, or via email. Once they click on it, the computer is infected and the criminal has access to reach all corners of your network or put whatever malware in place that they want.

You may notice that it sounds pretty similar to quid pro quo attacks, but rather than offering a service or pretending to be an I.T. support technician who needs to install critical software updates, they'll tell the victim that they won something, some terrific prize they can't resist—perhaps a catered event for the entire office, or tickets to a sporting event or some really cool and popular show nearby.

They can be as in-your-face as the *"You're the 1000th visitor!"* pop-ups you get on random, disreputable sites (consider blocking certain websites on your network to prevent employees from surfing the Internet instead of focusing on their jobs, anyway) to much more targeted attempts to break into your system.

ANATOMY OF A BAITING ATTACK

When baiting attacks specifically target certain employees, they can very effectively draw in those individuals. Criminals do their research on workers who, for example, have insufficient protections on their social media pages and thus make it easy to see where they are and what they like. This is another instance where effective social media management and extensive workforce training can save the whole business from becoming a victim.

Baiting can happen to anyone and at any time. Keep your security settings high, add protection to your websites and social media platforms without sacrificing your brand exposure online. Make sure employees don't check their personal pages on the clock. This doesn't just waste your time and productivity but opens the entire business up to more and more opportunities for a breach.

Consider a study conducted by the University of Illinois, which resulted in some startling discoveries. They put 300 USB drives around the campus. Of those, 48% were found and plugged into computers within minutes, and very few of the "victims" bothered scanning it first to make sure it was safe to connect to an internal network!

For fairly simple bait using a physical drive, it was overwhelmingly effective. Imagine how much damage a cybercriminal could do if they first researched a particular target through their, or the company's social media postings and actually designed bait tailored to get them wondering what they've won or what's inside.

Let's look into two case studies that show popular past baiting attacks and are also demonstrative of our evolving modern times.

Case Study: Stuxnet Computer Worm
WHAT WENT WRONG?

Unlike at the University of Illinois, this attack was no experiment. In 2010, a nuclear facility was about to experience the all-too-real consequences of what came to be known as the world's first digital weapon.

OK, it's a bit of an extreme example and sounds dramatic, but that doesn't make the moral of this story any less important. The Stuxnet computer worm was first discovered when it infected computers at Natanz nuclear facility in Iran over a decade ago.

So what happened? A double agent suspected to be part of a secret organization responsible for assassinating key figures, known as MEK, unleashed the attack on the system by simply loading a USB drive deep into the heart of the nuclear systems. The Stuxnet worm honed in on the SCADA systems, or supervisory control and data acquisition; critically important controls for the whole base. It then opened a backdoor for someone, safely off the premises by that point, to later remote in and re-enter the facility electronically and wreak havoc

Using a thumb drive instead of a remote cyberattack has certain drawbacks, but if they can pull it off like MEK did in this instance, criminals can specifically target weak spots in the network to do maximum damage. Physical break-ins like this work faster and in this case, the turned agent could stick around to guarantee it works since they were authorized for that level of access. Soon the USB file infected other areas of the network, although in the case of this computer worm, the shady organization didn't shut down the system like they easily could have. Instead, they took remote control of the network, a much more powerful position to have inside a nuclear plant.

For this Iran facility, the resulting damage was catastrophic. The infiltrators went after uranium enrichment centrifuges which enable the source material to work as fuel. The attackers effectively disabled key systems of the nuclear power program.

WHAT SHOULD HAVE HAPPENED?

While this case has few takeaways about what they could have done different, it does demonstrate what damage can be done if an infected external device like a USB drive is placed into a computer and the havoc it can wreak. If you see a suspicious USB lying on the ground, it could be anything from an experiment engineered to test whether you're undertrained and need to find the nearest cybersecurity seminar, or it could be a bad actor trying to gain access into your system to steal or hold your data hostage.

For nearly all businesses, the threat is real because untrained people will let their curiosity get the best of them. People love to find things or get things for free – it's why free is still the most effective word in marketing.

Whether it's stealing your data or holding it hostage with ransomware, take steps to prevent malware from infecting your system with rigorous security awareness training for all employees of the company, from the mail room to the board room.

And if you really are facing a terrorist mole infiltrating your organization to carry out step one of a complex plot for a secret group of assassins, consider installing some software on your network that automatically scans USBs for suspicious content before their files can be opened. Or better yet, implement a DLP (Data Loss Prevention) program that blocks USB access entirely!

Case Study: GoDaddy Holiday Bonus
WHAT WENT WRONG?

You've probably heard of GoDaddy.com before. In case you haven't, it's a massive internet company that sells internet domain name registrations and hosts websites for millions of businesses worldwide. With 20M customers all over the world and more than 7K employees, it seems like they're one of the companies who thrived during the COVID-19 pandemic since they offer services that businesses just can't do without and a lot of companies turned to online channels to survive and thrive during times of social distancing.

But that's not the case at all. In the early days of the pandemic, GoDaddy laid off 814 workers. They have also had data breaches in recent years that compromised the security of their, and their customers', information. Consider these facts to understand the following baiting email, in a tense atmosphere where both the GoDaddy executives and employees were feeling at-risk of another attempt and emotionally charged from a long and unprecedented year of lockdown.

In December 2020, GoDaddy employees received an email offering a $640 holiday bonus. Usually, the company enjoys a holiday party together to celebrate all their hard work during the year. So it wasn't very suspicious when they got an email announcing that the company had record-breaking profits this year and wanted to show their appreciation. With the coronavirus pandemic an ongoing threat, they couldn't safely host a big gathering like usual, so instead they offered a one-time holiday bonus. All employees had to do was fill out a form with their personal information.

Approximately 500 employees answered the call. But instead of a receiving a bonus, imagine their surprise, then, when just two days later they received another email from corporate—this time chastising them for their incredibly poor decision-making skills and assigning a mandatory security training course for everyone who filled out the original form.

You see, it was all just a test designed by GoDaddy executives to both test their employees level of awareness and teach them to be wary of baiting scams like this in the future. You can imagine the workers' reaction to this discovery: After mass outcry, the executives apologized for the insensitivity and promised to do better moving forward.

WHAT SHOULD HAVE HAPPENED?

Everyone has a lesson to learn here. It's important to remember that hackers do often disguise themselves as higher-ups on the corporate chain by changing a few barely noticeable aspects of the company name in the email address (*domain spoofing*). Just because something seems to come from corporate doesn't mean that you should automatically hand over whatever they ask.

The recipients of "holiday bonus" should have noticed a few discrepancies with the request. If it was a genuine offer, the company should not need to ask them for all their information, when that should be part of their professional files already. Secondly, if they were getting a bonus then it stands to reason that would get deposited through whatever method GoDaddy uses to send paychecks; whether direct deposit or a physical check, any holiday bonus should be easy to send out the same way.

Granted, the GoDaddy employees' only mistake here was believing an email sent from their boss but they would still benefit from security training (maybe don't make it mandatory as a punishment for answering a summons from the CEO, though).

The GoDaddy executives have some things to learn here too. For starters, it's not the best morale booster to trick your workers by promising them a lot of money, especially during a global pandemic when the average American is in a severe financial bind and could genuinely use the bonus. Especially given that they just laid off hundreds of workers a few months prior, employees won't be at the top of their game working in that kind of fear and uncertainty.

If you want to teach your employees to be smarter about cybersecurity, it's best to be honest and transparent about their mandatory training and offer incentives before punishment. When you do put them through cybersecurity exercises, it's better to teach them the tricks they need first before setting test emails loose during the holidays.

The overall idea came from a good place, though. GoDaddy has been targeted by cyberattacks in the past and have had data stolen as a result. Baiting attacks do happen too, especially surrounding the COVID-19 pandemic, promising tempting offers or information about the virus and vaccines. Be careful and teach your employees the proper way to approach all correspondence—just do it respectfully.

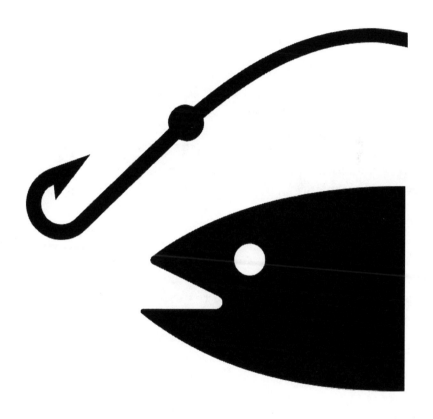

Chapter FOUR

An Introduction to Pretexting

Remember all those spam calls you tend to get throughout the year? Maybe they're telling you that your car registration is about to expire and need to confirm your identity. During midterm and presidential elections, you might experience an increase in political robocalls about candidates and voting. No matter what the scam is, they'll often rely on a faux position of authority to get the target to believe more completely in the lie.

Pretexting doesn't rely on outright fear, threats or tempting offers like exclusive business subsidies for COVID-19. Instead, this social engineering attack fosters a false sense of trust between the perpetrator and victim. They'll pretend to be some kind of authority figure to convince you to give over your personal information. All of those spam calls pretending to be from the bank or credit office, only to turn around and ask you to verify your information first, are making shoddy attempts at pretexting.

Criminals are more cunning than this now though, and pretext attempts are less obvious than an automated message from a machine.

ANATOMY OF A PRETEXT ATTACK

The lie to get your personal information or remotely break into your business's network is only the first part of the plan—the *pretext* part of the *pretext attack*. The *attack* comes next.

Once they're past your defenses, the cybercriminals go through your files to commit identity theft or commit secondary attacks, like releasing malware or ransomware onto your systems. The "pretext" refers to the way that they get past your security rather than what happens once they are inside. The trouble lies in what they can get into and away with when they have your personal information or access to your business network.

- According to Verizon's Data Breach Investigations Report, pretexting and phishing combined make up 93% of social engineering breaches each year.

- Their primary way of reaching out to prospective businesses is via email.

- The FBI found that business email compromise scams, which primarily target high-level executives, made up half of successful cybercrimes in that same year. On average, each successful scam resulted in an average loss of $75K.

Pretexting and phishing don't necessarily go hand-in-hand, but they often do which is why those two types of attacks make up nearly half of all cybersecurity attacks on businesses and other professional services. It's critical to be prepared.

Case Study: HP Computers
WHAT WENT WRONG?

Perhaps one of the most famous computer engineering companies also had one of the most well-known pretexting scandals. Though it happened over a decade ago, the details could easily happen today.

Hewlett-Packard, known more commonly as simply HP, experienced a crucial data leak in 2006. Journalists had somehow found out about their long-term plans for the business, which could of course prove disastrous for their competitive edge. Thus HP quickly rallied to try and find the source of the leak.

The board of directors, led by then-chairwoman Patricia Dunn, hired external security experts to assess who had told the press about their plans. Given the confidential nature of the information, they knew the leak had to come from someone high up in the organization with access to confidential strategic information.

Following that thread, management gave the investigators all the information they had on each of the board members—including their social security numbers, and everything and anything else they needed to impersonate those same members they were investigating.

That's right. The security forensics team turned around and impersonated both various board members and journalists, including those from powerhouses the Wall Street Journal and the New York Times. They then called around to the various phone companies requesting the call records for all their directors, and the bounty came in.

Ultimately, the leak was traced back to another board member. Dunn stepped down, replaced by the CEO of the company, for her part in the fiasco. Suddenly, pretexting went from obscurity to a prevalent cybersecurity threat that people began to take seriously.

Starting with Congress, although a previous law was already in place from the '90s that prohibited pretexting to gain access to financial information, the Telephone Records and Privacy Protection Act of 2006 expanded these protections to records held by telecom companies like those targeted by the criminals impersonating HP. Since then, the threat of pretexting attacks has grown with the internet.

WHAT SHOULD HAVE HAPPENED?

The increase in digitization across every industry, but particularly the professional services that hold onto very private information like call records. All of that is online and guarded by a select few—so those in charge of keeping that data safe need to remain vigilant so that they don't become victims, too. More and more often, criminals don't even have to call—now they can email or otherwise digitally reach out to request access to private information.

Take proper precautions and make sure your team knows to ask the right questions before giving out intensely private information. Criminals can and will impersonate others with high security clearance, and even companies that you've worked with closely before. Just because you're used to giving out crucial information to your computer company doesn't mean it's always them on the other line.

Those in high positions at your company are particularly prime targets for hackers and need to be extra wary. They should participate in cybersecurity awareness training courses, especially given that they have access to the most sensitive and important data in the company, giving them much more to lose if they become the source of a breach.

As for HP, they should never have given out such sensitive information as their board members' social security numbers to anyone, let alone an outside party. Treat your own security as cautiously as you do your customers. Would you give out your best patron's private information to an unknown third party? Then don't do it to your own board members, either.

Pretexting isn't so unknown anymore, and businesses now know that it's a very real threat to their privacy. Educate and take precautions not to fall victim to such a simple but clever ploy.

Case Study: Maria Butina and Anna Delvey
WHAT WENT WRONG?

Pretexts don't have to come from unknown sources online. People with charisma and the drive to scam you out of hard-earned money can get far by mocking up approachable social media pages and asking for the money straight-up.

Take the popular story of Maria Butina. She was part of the American political scene circa 2016, despite her ties to Russian intelligence. She worked for someone high up in the Russian Central Bank with the aim of infiltrating groups with influence in Congress. Arranging dinners in Washington D.C. and New York City, she made connections with pro-gun groups and others who sponsor politicians and thus have sway. She used profiles on Facebook, Twitter, Youtube, LinkedIn and other popular sites to make her backstory seem more believable. Eventually, she was found out and convicted of conspiracy in 2018. A year later, she was released and deported back to Moscow.

It's not just Butina. More people are leveraging their charisma to commit fraud. Consider Anna Delvey, also known as Anna Sorokin, who ravaged New York City out of hundreds of thousands of dollars. Using her pseudonym, Delvey, she targeted high rollers by pretending to be a German heiress who frequented popular spots for the New York elite. Her pretext for needing money and making these connections was to open a private club focused on art. With locations all over the world from L.A. to Hong Kong to London, she claimed needing a loan of approximately $22M to open her dream. It's a doable goal with the right interpersonal skills.

With Delvey's big spending and her tendency to stay in expensive places, her associates all assumed that she had the capital to back up her claims—but really, she financed the lifestyle by skipping out on massive bills or getting other people to pay off her debts, often to the tune of hundreds of thousands of dollars.

Delvey relied on forged bank documents and her own confidence to scam people out of nearly $275K total. This included airfare to places like Morocco, private jets and rooms at expensive hotels. Eventually, her lies came to light and Delvey was charged with attempted first-degree grand larceny, second- and third-degree grand larceny, and theft of services. She was sentenced in 2019.

WHAT SHOULD HAVE HAPPENED?

More and more frequently, social engineering attacks rely on social media as the entry point for scammers to establish a personal connection with their marks. But no matter how confident they seem and even if they flash a lot of physical cash, that doesn't mean these supposed socialites had any real money in the bank.

The moral of the story here is don't take anything at face value. Anyone can be a scammer, even when they look like you, so it's best not to invest if you can't *identify* that they are who they say they are. Scammers are smart and will ingratiate themselves into your circle by seeming to share your interests or social connections. Pretexting attacks don't always start at a computer screen.

Hence, be very wary of new acquaintances who just so happen to fall into hard situations when you're around. If they frequently promise to pay you back or ask you to help cover them for big expenses, just remember that pretexting can involve meeting people face-to-face and then using charisma and charm persuade you to help them. Always be on your guard, and never assume that all threats are cyberthreats. They can begin as subtly as an innocent conversation at a party.

This Photo by Unknown Author is licensed under CC BY-SA

Chapter FIVE

An Introduction to Man-in-the-Middle Attacks

As savvy as you've become at identifying small discrepancies that differentiate sneaky cyberattacks from genuine communication, verifying the identity of your correspondent sadly isn't enough.

Cyber criminals have developed clever ways to intercept and alter messages to influence their victims for their own purposes. This is known as a **man-in-the-middle attack**, where genuine messages are intercepted and potentially changed. This gives attackers unprecedented access to confidential information shared over your secure network.

If you share private and identifying information with a coworker, they'll see it. The message you're receiving from a valid source might have been altered by a bad actor already living in your network or email system. Man-in-the-middle attacks are a big reason why it's so important never to share confidential information by email or over the Internet. Even with all the proper precautions in place, you could still find yourself victimized even with everything appearing perfectly normal.

The two most common types of man-in-the-middle attacks are **eavesdropping** and **traffic modification**.

Eavesdropping refers to a data gathering technique used by criminals who have set up camp inside your communications network such as email and chat systems. They monitor all of your communications, building a database of critical, sensitive and/or confidential information over time that eventually allows them to break into your system or commit other cybercrimes.

The other is **traffic modification**, wherein they actually change the content of the messages to get their victims to do or say something that will get them their desired results..

Both are very dangerous, and you need to remain vigilant to avoid falling victim to either kind. According to Threat Intelligence Index 2018 compiled by IBM X-Force, man-in-the-middle attack encompassed 35% of attempted cybersecurity violations. Businesses like yours need to make sure you're not part of that 35%.

ANATOMY OF AN EAVESDROPPING ATTACK

The perils of eavesdropping, and what the hackers rely on is your lack of knowledge that the attack is even going on. By monitoring communications over a long period of time, they can accumulate nuggets of information that eventually build up to a serious cybersecurity risk. Since you're sending messages to people you know you can trust, you're more likely to include confidential information as it comes up in conversation. Over months of this, the bad actors can and will get personal and confidential information such as bank accounts, financials and trade secrets

Does that really sound impossible? Consider a CEO who emails her CFO the wiring instructions for a new deal they just signed, or the VP of Research and Development who send design specifications for a new product to marketing, of the controller emailing the sales manager their new credit card information for T&E expenses.

Cybercriminals can eavesdrop on your business in a number of ways. It can be as straightforward as breaching the network at a vulnerable place and implanting software in communication channels to collect all the data. The criminal can either listen in real-time or retrieve the information later.

Rather than hacking in directly, cybercriminals can alternatively send viruses to users in the network that implant the eavesdropping software automatically after it infiltrates the system. Since no data is immediately removed and there isn't ransomware holding your information hostage, you might not immediately recognize that your system has been compromised until the eavesdropper already have the information they were after.

Case Study: Alexa and Google Home
WHAT WENT WRONG?

Alexa and Google Home are wildly popular, and so are the concerns cybersecurity professionals have about them. These electronic devices have proliferated across the U.S. and around the world in the past several years as a convenient way for people to remember important dates and events on their schedules, prompt automatic reminders to complete tasks during certain points in the day, turn lights and TV on and perform everything else you'd imagine from an AI voice assistant. Despite all these upsides, users have always had concerns about their devices listening in when they're not supposed to. For smart speaker users, those worst fears came true in 2019.

Security Research Labs is a "hacking research collective and consulting think tank," according to their website. They investigated longstanding rumors into the security concerns most common with these voice assistants—rumors that have not gone unsubstantiated. An error was discovered in Amazon Alexa's coding, embedded in the calculator function. That let apps listen indefinitely to conversations after users were done with the devices. People thought that their smart assistants were turned off, but really they remained on and recording so they could report private conversations back to the creators of whatever app they were using.

Security Research Labs developed eight voice apps that were intentionally flawed to test the security of Echo and Nest devices. All eight passed the companies' review processes for third party apps, so they were allowed to connect to the system. Essentially the way these apps worked was that the developers inserted lines into the coding that the text-to speech AI couldn't pronounce. This would make the user think it was done recording, but secretly they would continue in the background without the owner's knowledge. Much later, after an extended silence that was also coded in, the app would say something like, "Your device needs a security update. Please say 'security update' followed by your password to start." So people believed that it was the Alexa or Home device itself, and would divulge the information freely.

Given that these devices are just complicated voice recorders connected to the internet, they're supposed to collect your voice and information and send them to servers at Amazon, Google or Apple—whichever company your device is from. However, they can also be sent to the third-party app creators to use for more malicious reasons.

After being notified about these privacy violations, Amazon and Google both released statements that boiled down to their steadfast dedication to consumer trust and confidentiality. They blocked the malicious apps and put up safeguards as well as additional steps to the vetting process, so that similar incidents would never happen again. Both companies also released preventative statements about how their voice assistants would never ask users to share private information aloud, so they know for the future not to trust apps that attempt to manipulate them in this way. Customers also have to opt in or out of accuracy review programs before their voices get sent to any outside parties.

WHAT SHOULD HAVE HAPPENED?

Amazon, Google and Apple have all had accusations against them for hiring independent contractors to listen to random conversations to, theoretically, improve the accuracy of their products. In reality, it demonstrates that your information might be unsafe even if it only goes to the servers where it's directed to go: Anyone could be listening on the other end, and you never know what their intentions are. This happens whether they hire independent contractors or not.

If your business relies on an AI voice assistant, it's unlikely you want to throw out all your technology and go back to taxing manual systems for reminders and information. Still, incorporate fair warnings into your training orientation programs to train your staff not to share private information like passcodes in range of their voice assistants. As we now know, never provide that same data directly to the apps either— even if the device wants you to. They shouldn't ask for that information over AI, and it's probably malware.

Case Study: Ring Doorbell
WHAT WENT WRONG?

As we develop new products to take advantage of advances in technology, developers also find unexpected, self-created problems. Amazon found this out the hard way just two years ago, when their Ring Doorbell Pro had a vulnerability exposed. Their users were susceptible to man-in-the-middle attacks.

Here's how the Ring works, for those who don't know. It's a doorbell with a camera attached, so users can record, play back and capture movement outside their doors. This Internet of Things (IoT) device comes with an associated app so users can view all the video on their phones. It sends this information over WiFi—leaving data exposed during setup and configuration.

Hackers learned how to target that timeframe. They would choose a victim and shoot false de-authentication messages to the Doorbell, thereby convincing the device that you were offline. At that point, the Ring would immediately go back into reconfiguration mode, because it's set up to notify immediately when it gets disconnected. Theoretically, this gives users a chance to fix problems immediately so they never have to worry what's going on beyond their front door.

But when hackers convince the doorbell it's offline, they can view your WiFi credentials as the Ring hands them over in plain text. Once they can view that information, the criminal could connect to the same network and then get into other devices on your network. Clearly this has immediate repercussions for businesses, but nobody wants their data so vulnerable. Whether they just keep intercepting and reading your information, or diving deeper to steal files, or watch you with your own cameras to see when as you come and go, this severe vulnerability had to go.

WHAT SHOULD HAVE HAPPENED?

It took a few months for Amazon to patch the vulnerability with an update, but they did come out with a solution later that same year, in 2019.

Unfortunately for the buyers, there's little they can do to prevent man-in-the-middle attacks on their IoT devices. Developers at Amazon can change how they address similar situations when they inevitably occur as we continue to push the boundaries of what's possible in tech. Respond swiftly and openly about what went wrong and how the company plans to fix it, even if the patch can't be made instantaneously. But you'll know to test for similar vulnerabilities before launching new devices or creating similar software in the future.

This situation with the Ring comes during a time that's filled with great unease toward IoT devices in general. As people have become more aware of how vulnerable it makes them, and how it can negatively affect their security, some have shied away from digitizing their lives like this. All you can do from the professional side is make sure you're always prioritizing your customers' security, no matter what cutting-edge tech you've got on the drawing board.

ANATOMY OF A TRAFFIC MODIFICATION ATTACK

It's a bit more complicated than simply reading confidential messages, but sophisticated cybercriminals can actually change the content of messages to more quickly gather relevant information from users. If you're not immediately giving up the private data that the criminal seeks, and they don't have the time or patience to wait, they can ask outright instead. Disguising themselves as a trusted coworker or friend is a great way to do it.

Modifying traffic attacks intercept a cache of data as it's being transmitted through the network, and then captures it. Cybercriminals can monitor and collect relevant data in the network, usually with different software programs that seek out particular things, like passcodes. Advanced software can gather that type of important information by counting keystrokes as messages, thereby reading everything that gets typed. Alternatively, certain programs give the attacker a chance to change the content of what's being sent.

These man-in-the-middle attacks show that it's important, not just to double-check that the sender information is accurate and coming from within the network, but also to assess the likelihood that the sender would really say whatever they're saying.

Case Study: Lenovo and Superfish
WHAT WENT WRONG?

Lenovo is a PC manufacturer and technology company. If you heard their name even six or seven years ago, though, you would have associated them much differently thanks to a scandal they had involving adware called Superfish.

People were already demonstrating some hesitancy and question toward Superfish in 2014, the year before the complaints started pouring in from concerned Lenovo customers. The adware was already under question when the real damage began. See, the Lenovo company had preinstalled Superfish on all of their machines before sending them out to their new owners. The intention was to push their preferred advertisements on search engines depending on what consumers looked for; this would directly benefit the company by drawing in more revenue from businesses looking to partner with them.

Lenovo used a self-signed certificate authority to insert pop-ups into various websites, like frequently-trafficked search engines, even if they are encrypted. You quickly see the problem: Hackers can just as easily break the encryption key, or copy it, and then Superfish has already done the work of cracking secure HTTPs without the browser noticing a thing. Their job is done, and then the hacker can simply intercept, steal or even change the real messages being passed through the machine.

Shortly after the complaints started pouring in, Microsoft released a Windows Defender update that could remove Superfish adware from devices. Although Lenovo technically offered users the option to opt out of the Superfish software upon startup, ultimately a group of State Attorneys General made a settlement deal in 2017 for Lenovo to pay out $3.5M for preinstalling adware on their computers. The money was to be distributed proportionally across several states with customers affected by the lawsuit.

Just one day after this judgment, the Federal Trade Commission ruled for Lenovo to get clear notice and consent for all their preinstalled software moving forward, and they agreed to perform regular security checks on them too. This particular settlement did not require Lenovo to pay any more than they had already.

WHAT SHOULD HAVE HAPPENED?

From Lenovo's perspective, there's an easy answer as to what they should have done differently: Don't install adware on your devices or include any other preinstallation software that can be (and later, thanks to Superfish, actually was) broken and bypassed by hackers to use for their criminal activities. Bypassing secure browsers to push ads may seem like a good sales technique, but it opens the door to all kinds of criminal activity.

While it seems straightforward to chastise the computer buyers, as well, for not reading the fine print more clearly and taking the out they were offered the first time they booted up the computer, it's not entirely realistic to expect people to do this. Anybody who's ever scrolled past a longwinded Terms & Conditions just to hit 'accept' at the bottom knows that people won't read all that jargon. Whatever courts may say about where the blame lies, most computer users simply won't read pages upon pages of legalese.

Instead, you can make your systems more secure with added encryption, security checks and software that sends false messages across the communication channels to make it harder for cybercriminals to track traffic patterns. Therefore, it's more difficult to decipher what's a real message they can intercept and change, and what's simply white noise to confuse hackers like them.

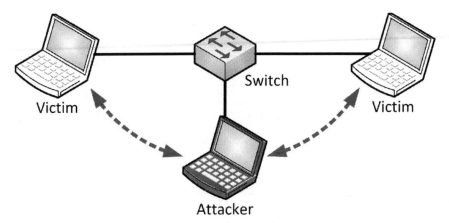
This Photo by Unknown Author is licensed under CC BY-SA

Chapter SIX

An Introduction to Cloud Vulnerability

Imagine this: You run a small business in a big city, where rent is high and space is in short supply. You process a lot of sensitive information for a huge customer base, but because of your employees, desks or other bulky office supplies, there isn't much space left in your tiny office to fit all of the servers and systems necessary to handle your company's processing and data storage needs.

Welcome to why the cloud has become overwhelmingly popular. Unlimited processing power and data storage capacity, as well as anywhere remote access to all your business data and applications, that can be as, or even more secure than on-premise, physical systems make cloud computing the go-to solution for many businesses today. It works faster and more efficiently no matter how many users you have or how much data you're working with, which gives businesses unlimited opportunity for growth without capital costs associated with infrastructure upgrades or refresh cycle.

You can set the privacy settings to give you secure access from anywhere there is Internet access, regardless of whether you're in the office, at home or on the road. This allows management to remotely run their business when they need to be away from their desks for any reason. This enables top executives to check on their business when they're away and thus makes it easier to manage multiple offices and all of your employees at the same time, increasing productivity and reducing down-time.

But as safe as the cloud is (when it has been set up properly), there are also dangers inherent in keeping your business and data behind a login screen away. The cloud may have arrived, but it doesn't come without its own inherent risks.

ANATOMY OF A CLOUD-BASED ATTACK

The beneficial aspects of relying on cloud-based security are also the sources of its greatest weaknesses. Given that you can access your data remotely, you run the risk of simple human error getting in the way. Have you ever tried to log onto your laptop while you're out traveling? The airport WiFi probably asked you how you wanted to connect—as though you're using your personal computer or with safety precautions in place for unsecured networks? Simply clicking the wrong choice could expose your data to outside threats.

Mostly, though, cloud vulnerability results in data breaches that can cause your business to lose massive amounts of data if you don't have additional backups in place. DDoS attacks, or distributed denial-of-service attacks, prevent you from accessing data and can significantly disrupt the flow of business until your team can solve the problem.

Cloud vulnerability matters more than ever before. Since the COVID-19 pandemic began, we've seen a massive shift toward nonphysical alternatives to facet of business' operations, and across all kinds of industries. The number of businesses offering cloud computing and cloud-based backup services doubled in 2020 according to Kaseya Business Benchmark Survey Report. Despite that, not nearly enough of these businesses are regularly testing their disaster response plans to properly respond in the event of a attack on their cloud systems.

As the world gets more and more digital, cyberhackers are also learning to exploit the very systems we're inventing to meet rising demand for remote security options. Technology will only advance; we need to step up and solve for the vulnerabilities before hackers can do serious damage.

Case Study: Uber
WHAT WENT WRONG?

Uber, the most well-known ridesharing apps on the market today, so much so it used as a verb in most cultures, was rocked by scandal a few years ago that demonstrated to many businesses and consumers alike that cloud vulnerability was a massive threat to privacy. It was one of the most explosive stories to reach the general public regarding this kind of security threat.

In October 2016, hackers compromised the private information of 57M Uber customers. The data stolen included their names, email addresses and phone numbers. The breach also affected Uber drivers to a lesser extent: 7M of them had data compromised, including over half a million of their drivers license numbers.

How did this theft come about? It was unnervingly easy. Two cybercriminals got into the GitHub coding website that Uber software engineers use to create and make the app run smoothly. Using log-in credentials gleaned from there, the hackers got into the web service that Uber uses for all their cloud computing and data storage. At that point all they had to do was email Uber executives with their demands and wait for the payout.

Uber complied with the $100K ransom demand that hackers gave, in exchange for deleting the stolen information and keeping quiet about the entire thing. Uber secured the data, closed the hole in the defenses, restricting the hackers' access to the information that they had previously exploited, and paid the ransom demand. The hackers, in turn, destroyed the stolen data.

Honorable thieves? No! Hackers know that if they don't keep their word, the chances the next victim will pay go down dramatically.

For a year, this security breach went unnoticed and unpunished. Once discovered, Uber made their excuses that no credit card or trip details were compromised, so there was no real harm done, but nonetheless they had a legal obligation to report the hack to appropriate authorities as well as the affected customers and drivers so they can take their own post-breach measures to protect their security. They have a right to know that someone accessed their personal information like that, regardless of if they lost funds.

Although the company has since strengthened their security posture, it's not the first time that Uber has been accused of breaching safety regulations in the interest of maintaining the illusion of business as usual. In fact, the previous CEO was ousted just months before the breach for putting the company at legal risk, too, in a lawsuit involving data security, disclosure and a settlement negotiation with the FTC. You wouldn't want your company getting a similar reputation for secrecy and undisclosed security issues, or risk legal action and fines.

Following the attack in October 2016, Chief Security Officer Joe Sullivan and another executive stepped down in response to the breach. They were sued for negligence in a class action lawsuit that went to arbitration in 2020. Uber publicly announced their intentions to take accountability for what happened, update their security measures, strengthen their security team and give free credit protection monitoring and identity theft protection to affected drivers. They've also agreed to perform security audits for the next two decades.

Notably, Uber refused to identify the cybercriminals although they were caught and ultimately plead guilty in 2019, to charges of computer hacking and conspiracy. Their charges actually included, not just the breach against Uber's private database, but a similar hack against Lynda.com (now known as LinkedIn Learning) as well.

WHAT SHOULD HAVE HAPPENED?

The breach of Uber's data and subsequent falling out is unprecedently large, thanks, in part to their extensive reach and popularity. Most cases like theirs don't receive as much national attention. Nonetheless, this data breach clearly demonstrates the dangers inherent in cloud computing, or any systems that can be accessed from the Internet.

In Uber's case, transparency and honesty would have been preferable to discretion. Businesses shouldn't use customers' and workers' data as leverage or playing cards in the larger game of cybersecurity. When they found the hack, they should have reported it to the proper authorities and disclosed to their users and drivers exactly what was going on so they could take their own precautions as they saw fit. Remember, hacks don't just affect you—they take a toll on your customers and employees, too.

Regular security checks also reduce the potential for issues long before they occur. We, as a society, are getting more and more reliant on cloud technology as its capabilities expand, security measures strengthen, and more and more businesses discover the benefits of cloud computing. As cloud technology advances, it's likely to integrate even more seamlessly into our day-to-day business.

Just because you can do more with your network doesn't mean there's inherent danger, nor does it mean your business is intrinsically safe. When attacks on the cloud do happen, follow appropriate the protocols to avoid more damage and legal trouble. Overall, though, it is most important to remember that reliance on cloud technology isn't going away. Rather, we need to take steps to keep our customers, employees and company data safe and secure before things go awry in the first place. Do regular security checks to guarantee your cloud systems stay secure and make certain to have, and test a disaster recovery plan in the event an attack is successful. It could save you from public scandal, a damaged reputation, and a legal trouble that might put your entire company at risk of survival.

Cloud technology is here to stay, and the vulnerabilities that come with it will likely expand as we get more and more reliant on it. Perform the necessary steps to make sure your business does not get exploited along the way.

Chapter SEVEN

An Introduction to Tailgating

Sometimes, the easiest way into a secured network is to follow someone in who has the rights and privileges to get in.

A **tailgating attack**, sometimes called a "piggybacking" attack, is just that: The criminal follows close behind a legitimate employee to gain access to private information that they couldn't reach without help. Have you ever been followed so closely behind on a highway that you're sure just tapping your brakes presents a risk of crashing? It's the same concept, except instead of a BMW, it's a hacker close on your tail.

It's one of the most prevalent threats in cybersecurity because it's so simple: Attackers don't need hacking know-how to do it. It can be an in-person, physical breach, like someone following a high-level employee past the doors to a secured area but is more likely a virtual trespasser following someone through a network backdoor.

Although it may seem like a straightforward method of entry, these attacks can cost businesses like yours hundreds of thousands of dollars in just one instance that's gone terribly wrong. Stay informed about how these attacks happen so you can make sure your business doesn't fall victim to a tailgating attack.

ANATOMY OF A TAILGATING ATTACK

Tailgating can happen digitally, or it can be a physical theft of the information that you store onsite.

When it happens through a computer, hackers will piggyback their way into a system that authorized users forgot to log out of. Say they somehow get remote access to your computer; if you're logged in, it's easy for them to click around for important, confidential information given that someone else has already gotten through the hard part of getting into the system itself.

When these attacks happen in-person, they follow a similar pattern. Common courtesy is a surprisingly strong motivator for victims; consider someone in a delivery getup with their hands full, calling out to hold the door. Most people will thoughtfully—or thoughtlessly, depending on how you look at it—let them through and think nothing of it. Just like that, your small business has been compromised thanks to people trying genuinely to do the polite thing.

Take, for example, the case of Colin Greenless.

Case Study: Colin Greenless
WHAT WENT WRONG?

Colin Greenless is a security consultant at Siemens Enterprise Communications, but his job is a little more unusual than those in most security firms. Greenless has a unique approach to cybersecurity: He infiltrated a FTSE-listed financial firm himself to show the holes in their network the most direct way he could. By exploiting them to show how easy it is, he could direct their attentions to the most vulnerable parts of the network with firsthand knowledge.

Here's how he did it: First, he chose the physical approach to getting inside. After some reconnaissance to get a sense of the security in the building, he felt confident that he understood the obstacles in front of him. Then Greenless just pretended to take a phone call, got in the elevator and mimed the floor he wanted to reach to helpful passersby. They let him right up.

Once he reached his intended floor, he walked into an empty office and set up his new work station right out in the open, where nobody looked twice. They do say that the best place to hide is in plain sight, and Greenless proved it.

Then he did the same thing in offices all over the building. Within twenty minutes he had access to high level information gleaned from files left out on employees' desks. His finds notably included a merger worth hundreds of millions of dollars. From the phone call scam to holding full cups of coffee in both hands, he just had to ask other employees to hold the door to gain access to all kinds of rooms, cabinets, desks and highly confidential information that had been left out in the open.

It gets worse.

He then faked a security audit to delve even deeper into the company. Considering they left multimillion-dollar contracts out in the open, you can only imagine what they had stored on their internal servers. He had the same idea. By pretending to be a security auditor, Greenless got this financial firm to give him all manner of private information about the network, just by asking directly for it. He invited another consultant from his own team into the building and got help analyzing the financial company's internal systems. Greenless quickly reached a first name basis with security guards in the firm too. He never hid his presence or where he was poking around, and they believed him largely because of his confidence. If you act like you're supposed to be there, a lot of people won't question it.

Greenless plugged his laptop directly into the server to tighten his hold on the company's secure data. As a business owner, you can just imagine what Greenless could obtain if he were to plug into the heart of your network. All that confidential information ripe for the plucking,

It gets worse.

Greenless found an internal phone directory with employees' names and information. He called twenty staff members who were listed there and continued his ruse of pretending to perform a security audit on the company. Given that he was on-premises and looking at their contact information, a lot of the employees believed him. After all, he had their personal information in front of him. Greenless asked these employees for their usernames and passwords "to help with his audit," and seventeen of them gave up the information freely.

All in all, it's a disturbingly simple ploy. It's a testament to just how flimsy all your internal safety procedures are when faced with a confident, charismatic, and full-handed individual like Greenless. At least he was doing it with the intention of exposing their weak points so that they could shore up their defenses in case of a real attack, but your business might not get so lucky.

WHAT SHOULD HAVE HAPPENED?

Granted, this is Colin Greenless's job. He knows what lies work to get him into the building and past otherwise secured areas. Regardless, it's a cybercriminal's full-time job to mess with your internal systems too. This case demonstrates just how easy it is to use tailgating tactics to gain unlawful access to highly secured, very expensive information and client contracts. Social engineering doesn't have to be digital to have a huge, negative impact on your network.

There are a few simple ways to lock down your physical security so they're guarded as safely as your digital systems. First of all, make sure your security personnel and your employees know not to give anyone access to restricted areas without the proper credentials. If someone shows up in a delivery uniform, boxes stacked in their hands and asked you to hold the door and take them to a secured floor, it's better to cause a little inconvenience than risk compromising the entire network if they're trying to break into that private area. They shouldn't be afraid to ask for identification or proof that someone is allowed to be in the building, especially if they're asking to enter a secured area.

Employees should keep their ID access cards on them or otherwise secured somewhere so no one can use it without their authorization. If you've got a Greenless situation unfolding for real, even having them out on a desk could prove risky. Remember, he gained access to a lot of unlocked offices that had private information lying out in the open on the desks. He could have easily taken a real ID access card to use on locked areas if his ruses didn't work. Always lock your workstation up when you walk away from the desk and don't leave private information lying around, even in the office.

Make sure everyone in your company understands the importance of cybersecurity awareness by holding training sessions that teach them the day-to-day precautions they should take to keep private information secured. In events like these, anyone can be victimized for the attacker's gain. Tailgaiting is a dangerous, in-person social engineering attack that you're better off preventing before the gears set in motion.

This Photo by Unknown Author is licensed under CC BY-SA

Chapter EIGHT

An Introduction to Water-holing

If cybercriminals want to target a specific person or group of people, they might reverse-engineer a trap designed to snare those specific victims. Getting close to them is easier when you already know what they like to do and where they tend to go online. This is called **water-holing**, also known as a cyberattack wherein the criminal gathers information about what websites their targets spend the most time on and then use those landing pages to infect the users' computer with malware or perform some other kind of secondary attack.

By learning their habits, a hacker can more easily create clever traps made just for your employees. It will be something that's likely to catch their eye, or at least that won't seem out of place in their inbox.

Water-holing attacks have hit several high-profile organizations in the past decade, demonstrating how critical protection is for any size business.

ANATOMY OF A WATER-HOLING ATTACK

The first step in a water-holing attack is for the cybercriminal to decide who or what company they want to target. Once they've picked a particular organization and somebody within it as the mark, then the attacker monitors and builds a profile on the victims. This might include details about which sites they frequent, when and how often, and how those landing pages might be effectively weaponized to pull off the water-holing attack.

The criminal finds weak points in the website and manipulates the code to capture the visitors' information and infect their computer with malware. From there, the cybercriminal has access to the targeted machines, and they can access the rest of the network easily from there.

For example, let's say you're a CEO who keeps their finances with Waters Bank. However, their security system isn't that good. One day, their website gets hacked; they overlay an invisible page on the actual website unnoticeable to the casual eye. When you enter your username and password, it secretly captures that information so they can break in any time on their own. Or maybe they compromise the site in a way that it captures the information when you click autofill; this is a more common problem than many people realize and why you should NEVER save your user names and passwords in your browser; auto-fill is a serious security flaw!

With water-holing in general, you can see a few immediate problems crop up. Since these are websites that your employees use regularly and know from experience have never harmed their computers before, they're more likely to trust them moving forward. They'll likely let their guard down, becoming less cautious as they use the website more and more and thus opening themselves up to a water-holing attack.

Learn how to detect and avoid this socially engineered cybersecurity threat so that the consequences don't happen to your business.

Case Study: Forbes.com
WHAT WENT WRONG?

In 2015, Forbes experienced a significant security breach aimed at targeting visitors to their website. More specifically, it targeted people visiting the defense and financial service-related pages.

Typically, when users visited the Forbes website, they'd be greeted with a "Thought of the Day" Flash widget, no matter which webpage they landed on. The guilty parties exploited a zero-day vulnerability, which just refers to newly found vulnerabilities in a software that can be easily taken advantage of by hackers because the developers haven't yet patched it, to implant spyware into both Internet Explorer and Adobe Flash Player. This alternative, malicious version of their widget loaded on every page of Forbes.com instead of the real thing, thus exposing anyone with a vulnerable computer to infection once they clicked it.

The breach, discovered by two cybersecurity firms, was the work of a Chinese-based hacking group believed to be responsible for several other similar attacks in the years previous. However, this marked the most popular website to experience this level of compromise up until that time. Since water-holing attacks are reverse-engineered traps laid for specific parties, they often occur on smaller websites that are sure to attract whatever specific community the hacker aims to exploit. In this case, however, despite Forbes's millions of hits per day, the attackers found a way to spy more heavily on their intended victims—those firms in the defense and financial service sectors.

When the spyware was discovered, Microsoft and Adobe both released updates to fix the vulnerabilities in their security shortly thereafter.

WHAT SHOULD HAVE HAPPENED?

In this case, both the affected software companies released updates as soon as the issue was discovered—the safe thing to do. As soon as you discover a zero-day vulnerability, you should report it so the exposure can get fixed with patches or workarounds immediately. It's best to also notify users, in the interest of full transparency for anyone who was affected.

This Forbes.com case also serves as a cautionary tale about why it's so important to undergo proper cybersecurity checks, also known as application penetration tests, before releasing software, especially when contracted with high-trafficked companies like Forbes.

Case Study: United States Department of Labor
WHAT WENT WRONG?

Before the incident with Forbes, there was a breach of a highly trafficked government website: The U.S. Department of Labor.

In 2013, the website found itself hosting malware that redirected visitors to a second website, which then infected their computers with Poison Ivy, a remote access trojan also known as a RAT.

What is a RAT? It's malware that secretly downloads onto a computer and then opens up a backdoor which the hacker can later enter to remotely take control over the machine. Poison Ivy is notoriously tricky type of RAT, and this version was updated to a point where it was made even more difficult to track than it had ever been before. Coupled with the fact that the Department of Labor would naturally be considered a trustworthy site, and you see why hackers chose this website to use as a watering-hole. People aren't usually suspicious of government websites and are thus willing to give up private information to them.

Multiple security firms discovered the Poison Ivy malware, similar versions of which were also found targeting other government pages related to nuclear information. They were targeting verified, informative pages with high-level information as a ruse, demonstrating the necessity of widespread cybersecurity software to protect computers from this type of secretive sabotage. It can go up as high as government-controlled websites.

When they discovered the Poison Ivy RAT, the Department of Labor shored up their virtual security and ridded themselves of the trojan.

WHAT SHOULD HAVE HAPPENED?

In this case, it's true what they say: The best offense is a good defense. Ensuring that your cybersecurity software is up to date prevents malware from getting into your website code in the first place, dangerous infections that could open your business up to even more damaging cyberattacks in the future.

Given that watering-holing attacks capitalize on the trust that targets have in their most frequently visited websites, potential victims cannot trust sites even if they've been vetted, seem like they should be trustworthy or if they've gone there many times before.

Develop a strong cybersecurity defense system and run regular maintenance checks, as well as updates, to make sure your computers are always working with the best that I.T. has to offer.

This Photo by Unknown Author is licensed under CC BY-NC-ND

Chapter NINE

An Introduction to Spoofing

Technology pervades everyday life; as your employees learn about how to protect themselves and adapt how they work to the realities of the threats that exist, cybercriminals too must adapt their playbook, finding more and more clever ways to trick their targets.

One of the better tools in the hackers toolbelt for reeling in unsuspecting victims is spoofing.

A **spoofing attack** is what it's called when cybercriminals send an email or other communication that appears to be from someone else, someone the intended victim trusts. The hacker leverages that trust to have the victim click on the infected link or file contained in the message.

Once the victim clicks the link or downloads the file, malware is introduced into their computer and onto the network.

And why would this be a great hacking tool, after all the victim believes it is coming from someone or some entity they trust. And because they trust the source, they are more likely to click on an infected link and accidentally download a virus containing ransomware.

Consider these statistics:

- In 2018, the Center for Applied Internet Data Analysis found that there are approximately 30K spoofing attacks happening every day.

- Spoofing can cost businesses hundreds of thousands of dollars per attack.

- More than half of businesses, on average, have insufficient Sender Policy Framework (SPF) and DomainKeys Identified Mail (DKIM) protecting their email—in other words, they're susceptible to spoofed emails *(Source: BitSight Insights Global View, 2018)*.

- Cybersecurity firm Agari found that in the first half of 2020, 68% of spoofing attacks used fake display names to trick their victims.

The three most common types of spoofing attacks to be aware of are IP, ARP and DNS.

Anatomy of an IP Spoofing Attack

IP spoofing sounds exactly like what it is: A cybercriminal falsifies their IP address to send network packets to other computers.

What does all that techno-babble mean? Let's get a few definitions out of the way.

An **IP address**, or Internet Protocol address, is a specific number assigned to each device on the internet, or any network for that matter. No two devices can have the same IP address on the same network because they identify and locate you. The IP address is used to route traffic to and from you; when you visit a website, the site knows where you came from and can send the page you are requesting back to you since it knows where you are located – at that IP address!

A **packet** carries information and data to and from a recipient or sender in the same network.

In layman's terms, IP spoofing lies about where the criminals' emails came from so that they can pretend to be valid users on the network or trusted source you can rely on. When you trust that they are who they say they are or come from an address you recognize you are more likely to trust what they send. This also makes you more vulnerable to malware and infected links included in these messages.

Case Study: GitHub
WHAT WENT WRONG?

On February 28, 2018, the largest denial-of-service attack recorded to date at the time hit the internet. GitHub, an open-source software developer site used by over 65M developers, was hacked via IP spoofing.

GitHub runs its memory through what's known as a **memcaching** system, often used by websites with high amounts of traffic aimed at high performance. By spoofing GitHub's IP address, the hackers were able to spam their servers with false traffic. That's where the **distributed denial-of-service**, or DDOS attack. They bombarded the servers with false traffic until they couldn't handle it anymore. With 50x its usual incoming traffic, the website went offline for nearly ten minutes.

As soon as they noticed the issue, GitHub called in outside help to reroute their information through a secure system that could remove and block the bad data. While they were working on pushing out the intruders, visitors experienced fuzzy and insecure connections. After nearly ten minutes, the cybercriminals finally called off the attack and the website was back online and safely in GitHub's own hands.

WHAT SHOULD HAVE HAPPENED?

When it comes to securing your web traffic, service providers need to set up firewalls that filter spoofed IP addresses and emails, and verify or block access to their servers pending proper authentication. Use appropriate encryption protocols to guarantee safe and secure traffic to your sites will reduce your risk of becoming a DDOS victim. Shoring up the security on your platform can prevent your technology from being weaponized against someone else, too.

In GitHub's case, they actually exhibited a pretty good reaction given how many companies rely on the platform. They noticed and handled the problem swiftly, and although it was the biggest DDOS attack to date, they reacted as quickly as they could. Then, after the hackers retreated, the website continued to look into what had happened so they could see what preemptive defenses would protect them against a similar attack in the future.

Despite the risks, plenty of businesses don't have procedures in place that ensure their device's firmware is up to date or their operating systems are fully patched. It is imperative that you update the firmware, software and operating systems, and their security settings, on your systems and IoT devices regularly—especially for older technologies that are not prepared to handle onslaughts driven by modern hacking techniques.

Everyday users need to be wary too, especially employees who log onto their IoT devices on company WiFi, thereby potentially compromising the entire organization. It often happens where hackers can get into a system by using the unsecured phones and devices connected to the same network. That's why in general, people should:

- Make sure the websites they visit are preceded by HTTPS:// and the padlock symbol beside the URL, indicating a secure connection.

- Avoid unsecure networks and don't send private information over public WiFi, like at the airport or a coffee shop, because you run a higher risk of compromise.

- Be wary of emails that ask for personal information, card data and donations, even from trusted senders. You should not send private information by email, anyway. If you ever *have* to then make sure you encrypt it.

- Don't ever leave default passwords in place. The first item on your to-do list when you buy an Internet-connected device is change the password. And use passwords that are difficult to guess, especially by probing malware.

The more connected we are, the bigger the attack surface becomes – all the ways in which cybercriminals can gain access to our systems and information. Cybercriminals, unfortunately have unprecedented ideas about how to compromise your servers and block access to crucial information.

Learn more about strong passwords and what other things you can do to button up your security in the second installment of our series: REVEALED! The Secrets of Protecting Yourself From Cyber-Criminals.

Anatomy of an ARP Spoofing Attack

Sometimes, cybercriminals don't have to fake a valid IP address. With ARP spoofing, they can connect their MAC address to a valid computer on your network, thus convincing the security systems that the intruder also belongs there. They're linked in a way that suggests they're authorized by association.

A **MAC address**, or a media access control address, identifies users through a unique physical address, rather than the logical address designated by IP. The MAC address may differ thanks to certain operational functions, giving criminals a unique avenue to exploit.

In an **ARP spoofing attack**, cybercriminals intercept their address resolution protocol requests and send false data in their stead, ultimately linking their MAC address to a legitimate IP on the network.

Case Study: Metasploit Project
WHAT WENT WRONG?

ARP tells networks where to send the traffic that it receives; which computer or other device connected to the network needs to be redirected to whatever website they're looking for. ARP spoofing attacks redirect traffic intended for a website or server resource to somewhere other than their intended destination. Then they can steal data meant for the *real* location or website.

This is what happened to HD Moore's Metasploit Project in the summer of 2008. Usually, this is a cybersecurity site with information and testing about various issues affecting the computer industry.

A group called Sunwear rerouted all their traffic to a page other than the intended homepage for the Metasploit Project, a landing page which declared that they were responsible for the ARP spoofing attack. This hijacking also came at the price of denial-of-service attacks on many services on the Metasploit website.

After HD Moore ascertained that the site itself worked fine, he determined that it was a network-based attack that went beyond his server. Ultimately, the ARP spoofing went all the way to the VLAN, or virtual LAN, that connected Metasploit to a subnetwork of many sites grouped on the same network. That's why it was able to redirect all traffic headed for the website without affecting the site itself when viewed by the creator. The misdirect only happened on outside parties' side.

The ARP spoofing attack affected over 200 other websites on the VLAN too. HD Moore fixed it on his individual network by hardcoding his router MAC to the ARP, but the other affected sites would have to do the same to get rid of the overarching problem.

WHAT SHOULD HAVE HAPPENED?

Hardcoding the MAC address of the system running the Metasploit Project website in the first place would have prevented such an ARP spoofing attack from happening. As always, preventative action is better than learning the hard way.

Regardless, it's helpful to be able to quickly identify when these attacks happen. Denial-of-service, sessional hijacking (a major concern when operating under public WiFi) and man-in-the-middle attacks tend to happen after a breach that starts with ARP spoofing. The hackers can disrupt, manipulate or access your data through vulnerabilities in the network while directing traffic somewhere else.

VPNs, or virtual private networks, will encrypt your connection to the ISP so hackers can't track your online activity. This is especially helpful when working on-the-go or when you're on public WiFi. However, this can slow down your internet significantly, so it's a trade-off that can become more frustrating than it's worth if you don't get the right one. Paid VPNs, of course, work faster than free alternatives.

Static ARP entry creates one permanent communication channel between two hosts in the cache that often talk back and forth. This creates one more barrier that hackers have to break through in order to gain access to your company's private data and web traffic.

As always, beefing up your cybersecurity helps. Installing detection tools and preventative software is always the best way to avoid a serious spoofing attack from affecting the flow of business. Talk with your IT provider to see what concrete measures you can take against ARP spoofing attacks.

Anatomy of a DNS Spoofing Attack

Domain name spoofing, or DNS spoofing, catches and redirects traffic to a malicious website by using a fake domain—as the name suggests. By rerouting intended web traffic from a real site to a different IP, the change up passes under the radar; visitors are less likely to notice the change and thus let down their guard, and expose themselves to more danger as the fraudulent site is typically designed to look very similar, if not identical to the real site.

Since the fraudulent sites are typically designed to pass as whatever real website the users wanted to access, people are less likely to notice something is wrong and end up giving away crucial, private information without thinking because they've done it before. They'll fill out their username, give up confidential information or even upload financial information or trade secrets freely, believing it to be their usual secure site.

DNS spoofing attacks can potentially mean that your employees hand their passwords and private data right to malicious actors without even knowing that something is amiss, until their files start disappearing or they're a victim of ransomware. You need strong cybersecurity safeguards like network level monitoring tools and secure DNS so you can catch or prevent DNS spoofs before they start wreaking havoc.

Let's examine some situations where victims weren't careful enough—and it cost them.

Case Study: Malaysia Airlines
WHAT WENT WRONG?

In 2015, Malaysia Airlines (MAS) passengers found that they couldn't get onto the official company website anymore. They started getting reports of a hack, but despite this, MAS checked their servers and found that everything was completely fine on their end.

That's because the root cause of the problem was neither their servers nor their hardware. The real issue lay deeper: DNS cache poisoning, a cyberattack where criminals alter the cached IP address in the DNS server, and thus they can redirect all the intended traffic for the MAS site to their own webpages. From the company's side, everything looked just fine. For their users, not so much.

The hack was traced back to Lizard Squad, a known cybercriminal group who proved hard to catch because of their technological savvy and international reach and membership. They openly bragged about the breach on Twitter, at which point MAS admitted publicly that their DNS had been hijacked. Nonetheless, they assured customers that all their data was secured since the hack wasn't directly on the website, but instead on traffic that it received. Lizard Squad disagreed, and threatened to leak private airline data.

Thankfully for their customers and their reputation, that never came to pass. Malaysia Airlines reported the breach to the appropriate authorities, both Cybersecurity Malaysia and the Ministry of Transport. The company resolved the issue privately with their service provider and recovered their system to the fullest extent.

WHAT SHOULD HAVE HAPPENED?

Malaysia Airlines actually did a lot of things right here. They were transparent and honest with its users, which builds trust and maintains their reputation even after a breach like this, which typically hurts their credibility with some people. Customers want to know exactly what's going on with their data, good or bad, and they especially don't like to feel misled by the very companies that they entrusted to keep it all safe.

Even when things go wrong, Businesses should notify everyone who was potentially affected by the breach so they can be with you every step of the way to getting it fixed. Honesty and transparency can't take back a mistake, but it demonstrates good faith and a commitment to doing right by them through what comes next.

Preemptively reinforce the security measures on your DNS servers to avoid having your data compromised in the first place. Encryption certifications for your domain names also help prevent DNS spoofing attacks moving forward. Better cybersecurity stands between your business and serious damage down the line.

Chapter TEN

An Introduction to Smart Contract Hacking

Everyday technology becomes a deeper, more intimate part of our lives. We are always striving for ways to do it faster, easier, more accurately; it is who we are as human beings, and we see evidence of this all around us. When is the last time you walked down the street without seeing anyone holding a smartphone? That's not an indictment of technology, whose interconnectedness has sparked all kinds of incredible inventions, globalized initiatives and advances not possible otherwise, thanks to these unforetold capabilities. Now, we rely on technology for so much more than we ever imagined when the internet was first invented. Business has become digitized down to the agreements that companies make with each other.

Smart contracts are a way to securely make and accept transactions with other users over the Internet, with all the security and binding legal power of a physical contract. They rely on blockchains, without third parties. You can program whatever stipulations you want into the contracts because of their self-executing codes, including autopayment on a regular schedule, depositing funds only upon receipt of service, to whatever other commands you want to program directly into the smart contract.

Unfortunately, for all that this has done for the future of business partnerships, this has opened up a wealth of new possibilities for hackers, too.

ANATOMY OF A SMART CONTRACT HACKING ATTACK

In the world of information technology, smart contract hacking is still relatively new. That means that there are plenty of bugs and vulnerabilities to exploit that haven't yet been fully exposed and eradicated from the practice of executing smart contracts. In the past decade, hackers have stolen $1.2B in Bitcoin and Ether, their rival company (*Source: Autonomous Research LLP, 2018*). As cryptocurrency gains prominence, the target on its back gets bigger too.

Since smart contracts are *authenticated* and pre-approved, they guarantee follow-through on both ends. An impartial, digitized third party executes the contract, so there's no chance of either side reneging on their ends of the deal. It all happens automatically, without interventions of any kind. However, this also gives cybercriminals the opportunity to hack in and manipulate the code to execute actions on their behalf instead.

Blockchains are digital ledgers for transactions that give everyone on the same network equal access to that information. By altering code in the blockchain, hackers can force the contract to execute specific actions before they're meant to happen, or exploit vulnerabilities in the code to access the money in the ledger and take it for themselves. As long as money is connected to technology, some hacker somewhere will want to exploit it for their financial gain.

Now that there's cryptocurrency like Bitcoin or Ether involved, cybercriminals are learning how to exploit technology to steal that too.

Case Study: Parity
WHAT WENT WRONG?

Parity Technologies is a blockchain engineering company that discovered the perils of smart contract hacking back in 2017.

Parity Wallet, a digital storage space for cryptocurrency, was initially developed so that it required multiple private keys to approve the transfer of Ether. These **multi-signature wallets** seemed like a good option for smart contracts to rely on, because of the increased security you get by needing multiple approved signatures to execute the code. One-side couldn't exploit it for their gain without the other's consent.

Enter an anonymous hacker. They stole 150K Ethers, or approximately $30M at the time, by exploiting two functions in the smart contract library. The first, **delegatecall** function, triggers one contract to call on the other to execute its code. The **fallback function** is externally visible and occurs when the contract calls on a nonexistent function, so they fall back on this default instead. The hacker wasn't identified.

Just a few months later, someone by the handle devops199 accidentally exploited another issue in Parity's multi-signature wallets. They found a bug in the code. They called on an initialization function that assigned them the owner, who could also cause the code to self-destruct. This prevented the blockchains from transferring any cryptocurrency and ultimately froze 513,774.16 Ether from nearly 600 wallets.

WHAT SHOULD HAVE HAPPENED?

As far as smart contract users are concerned, just be more wary of what smart contracts you're using to send and receive money digitally. Nearly half of smart contracts written in Solidity, the coding language exploited in the 2017 Parity hacks, have proven vulnerable to hackers. Knowing who, and how, a smart contract service elects to execute your cryptocurrency transfers should be part of your decision-making process when choosing a platform.

Overall, be careful before entrusting valuable assets or money to a digital service in the first place. Where there's a remote network, there are vulnerabilities for cybercriminals to exploit—and some are much more likely to get attacked than others, especially when there is a lot of money at stake. With modern technology, especially of the kind that's still relatively new to the world of cybersecurity, the road to a secure future for smart contracts is littered with land mines.

Case Study: Block.one
WHAT WENT WRONG?

The invention of cryptocurrency and its subsequent explosion of success has caused other companies to want to get in on what Ethereum and Bitcoin have. Success and financial gain are strong motivators for invention. That's exactly what Block.one did when they launched their own blockchain-based cryptocurrency, EOS.

Given how successful smart contracts have been in appropriately distributing funds to professionals after a legitimate exchange, it's easy to see how this technology has revolutionized online gambling too. It makes sense: Smart contracts will go wherever large transfers of money go, and virtual gambling is the perfect storm for cryptocurrency holders and criminals alike. Enter DEOS Games, an online gambling game that uses Block.one EOS.

The rules are straightforward: Players send EOS to their smart contracts that handle bets on their lotto, blackjack or roulette games. The smart contract automatically sends users their winnings if and when they win, so the game remains fair—the website even boasts that the house doesn't have an advantage. Theoretically, digitization should have taken all of the uncertainty out of the process.

In 2018, a hacker going by the handle runningsnail used their own code to disrupt the normal processes of the EOS blockchain and score way more from the betting company than they should have rightfully earned. It started with just a few small, discrete transactions that didn't raise red flags, then slowly grew to bigger payouts. Runningsnail would transfer 10 EOS to the game, and the code would deposit back way more than he put in. Over the course of twenty-four "bets," the hacker successfully stole more than $24K in EOS.

The day following his massive winnings, DEOS Games released a statement about the hack exploiting their system, how they spent over six hours developing patches for the issue, and overall called it a good "stress test" for them to improve their smart contracts moving forward. All in all, that's a pretty good PR strategy to ease their customers' minds and encourage future investments.

WHAT SHOULD HAVE HAPPENED?

When it comes to stress tests, companies do better by rigorously challenging their software – called an application penetration test – *before* it costs them tens of thousands of dollars. Also consider hiring an external user to audit the program and make sure it's as secure as you think it is before launching the service; sometimes all it takes is a fresh eye to catch a bug that could later bring your entire smart contract to its knees.

That being said, transparency remains the best option when something like this does happen. Your reputation may take a hit, but users' trust in you doesn't have to. Show that you're keeping their security in mind moving forward, and a lot of them will happily stick with your services even after a hit to your reputation.

Get to know common smart contract vulnerabilities so you can preemptively mitigate issues before your program launches. Don't let someone steal thousands of dollars in cryptocurrency before you address a major flaw in the code. Build better smart contracts from the start to avoid something like this happening to you.

This Photo by Unknown Author is licensed under CC BY-SA

Chapter ELEVEN

An Introduction to Rogue Scanner Hacking

After reading this far, you might be feeling concerned right about now that malware is around every corner. But cybercriminals have even found a way to spread computer viruses by capitalizing on the *fear* of computer viruses. It's called **rogue scanner hacking**.

This particular threat goes by several names: Rogueware, fake scanners, fake AV. All of them refer to a particular kind of attack where malware disguises itself as antivirus software to trick people into installing the rogue scanner onto their system, thinking it's going to get rid of malicious files. That's where the havoc begins.

Rogueware first conclusively popped up in 2003 with a fake scanner known as Spy Wiper. It could do a lot of damage including changing your settings, introducing relentless ads and watching everything that your computer did. Although it's evolved in the two decades since its first known infection, a lot of the core capabilities remain the same.

Rogue scanners are one of the two biggest types of **scareware**, which is deception software that bombards you with a variety of threats and alerts insisting you download a software to fix the issue. The software is actually malware unleashed on your system. The other main category of scareware to watch out for is ransomware.

There are many different kinds of rogueware out there waiting to trap you, from pop-up ads to fake threat notifications, potentially increasing how many victims they catch in their net. Some even try and play on your emotions to get you to buy into their scam, by promising that a portion of their subscription sales go to charity or other incentives like that. Don't underestimate how badly cybercriminals want your credit card number.

Now, none of this is meant to make you worry that deceptive practices and dangerous threats lurk in every digital shadow. It's a wake-up call for some, a warning for others but overall remain constantly vigilant. While plenty of antivirus software will do exactly what they say they will do, keep an eye out for those that aren't all they seem. Here's what you want to recognize.

ANATOMY OF A ROGUE SCANNER ATTACK

Typically, with rogue scanner hacking, pop-ups and other messages tell victims that their systems have been infected by a virus. Then they're urged to buy fake security software to remove the so-called threat. The malicious actors pretend to be AV scanners, hence the name "rogue scanner," but they can actually block your real antivirus software from identifying and destroying them once they're inside your system. This is on top of the actual damage they do to your files, money and information.

Rogue scanner hackers are smart. They can inject malicious code into a legitimate site so seemingly trusted sources are the ones generating pop-ups offering the bogus security scan. Since users have visited these websites before and had no trouble, they're more likely to trust these phony notifications. Hackers can also use real third-party ad platforms to put "malvertising" on those authentic websites. In these cases, the rogueware downloads when victims click the ads for the fake security software.

Given that they're designed to look like genuine and even popular antivirus software, it's a clever form of social engineering to get more downloads.

Think you're immune to the false promotions like these? Check out these real-world examples showing how easy, and how dangerous, it is to fall into the malicious trap of a rogue scanner hacker.

Case Study: DarkAngle
WHAT WENT WRONG?

Panda Cloud Antivirus is a real security software built by PandaLabs. In 2012, downloads soared for a scanner whose icon looked exactly like Panda Cloud Antivirus. Unfortunately for the users who installed it, that's where their similarities ended.

Free download sites combined with spam messages spread this trojan masquerading as Panda Cloud. In reality, the culprit was a rogue scanner called DarkAngle. Once victims downloaded the software, it logged every keystroke that they typed and sent it to a remote server. It was also set to automatically load on computer startup, guaranteeing that the rogueware caught everything these people did on their computers.

What's worse, DarkAngle came equipped with the features necessary to sneak beneath the radar of legitimate antivirus software. Upon download, it added 20MB of junk data to the machines, which prevents some legitimate antivirus software from scanning the file since malware isn't usually so large. They had other defense tactics against detection as well, including shutting down certain processes to make the systems more vulnerable to additional malware and cybersecurity threats.

Given that the trojan also installed new malware onto the machine, this was a critical component of the plan.

The DarkAngle rogueware touched all corners of each system, too. It accessed webcams and microphones so that it could send audio and video recordings back to the remote server in addition to the other data it stole. And once you've got DarkAngle on your system, you need real AV software to get it back out.

Surprisingly, this is not the first time that Panda Cloud Antivirus has had rogueware impersonate their brand. In the past, they've had fake AV scanners mimic their software that falsified scans, pretended to find malware and requested you buy their software to "get it out" of the system. The worst part was that if people didn't buy, they would still periodically get pop-ups with similar frightening messages when they tried to run nearly any program on the machine. And it's not just malvertisements popping up, because it would seem suspicious if you were getting random ads all the time, especially when you weren't online. These look like real system security alerts, mimicking their language and content just like DarkAngle copied Panda Cloud Antivirus's icon. It's a very effective scareware tactic.

Rogue scanners aren't just trouble; they're an enormous pain to deal with and get rid of, too.

WHAT SHOULD HAVE HAPPENED?

The best way to avoid fake scanner AVs is to learn the signs that the software in front of you isn't as advertised. Research and get to know the legitimate vendors and their software so you can tell the difference between helpful scans and rogueware scams, and always buy directly from the developer's website or verified social media to avoid downloading something like DarkAngle instead of Panda Cloud as expected.

If you do fall victim to rogueware, it's not the end of the world. Don't put a credit card down for AV scanners until you're positive it's real, even if you accidentally click on the link and you've been plagued by fake pop-ups like the DarkAngle victims were. Instead, get verified antivirus software and run it to cleanse your system entirely. A lot of the time, you can just close the browser tab that you accidentally opened and that's enough to prevent rogue scanners from bothering you any further.

Knowing the signs of a rogueware attack is the first, useful step toward preventing real damage to your system. Given that rogue scanners can change your settings, randomly crash your computer and block real antivirus software, on top of how it's so difficult to get rid of, then you need to act fast if you've been hit by one of these social engineering attacks.

There are versions of popular antivirus software that have come out in more recent years which are designed specifically to spot and eliminate rogueware. The real Panda Cloud AV developed such technology, but you can also find those capabilities with Norton and other brands as well. When it comes to protecting your system from fake scanners that open you up to additional attacks, it's better to build up your defenses in advance rather than deal with the potential fallout from a hack like that.

The Debate Over Rogue Scanners' Threat Level

There's been some discussions over the past several years about whether rogue scanners are as dangerous as they've been made out to be or if the risk is shrinking that you'll fall victim to this kind of attack.

Some cybersecurity experts believe that rogue scanners are actually on the decline. When it comes to scareware, they think hackers are relying more and more frequently on old fashioned social engineering tactics instead, then followed up by ransomware. Perhaps they just get more money with ransomware; whatever the reason, it's possible that rogueware really is on the decline. Yet despite the talk of its demise, it remains unclear.

There's also a possibility that people are simply becoming savvy to the tricks typically involved in rogue scanner attacks; or perhaps they are simply becoming more comfortable and more familiar with their own antivirus software; the better you understand how it works, the better you'll get at spotting anomalies that indicate a scam.

Whatever the reason, it's possible that rogue scanners are becoming less of a threat to businesses. On the other hand, conflicting opinions hold that they're just as prevalent, and just as dangerous, as they have been for years. Considering the damage that fake AV scanners can cause on your system when there is a breach, it seems, like all other threat vectors it's better to be safe than sorry when it comes to social engineering tactics.

Remember, scareware is designed to prey on your need for cybersecurity. They offer fixes that you didn't know you needed and rely on your lack of know-how and your fear. Don't let anxiety get the best of you; approach cybersecurity solutions with the same caution and vetting you would were you buying a car and you'll be just fine.

In general, anything that offers free services, asks to scan deep into your computer or wants payment up front is probably too good to be true. Scareware can do a lot of damage to a company, so whether it's becoming a more popular tactic for hackers or experiencing a downturn, it's best to be prepared and secure your business either way.

This Photo by Unknown Author is licensed under CC BY

Chapter TWELVE

An Introduction to AI Enhanced Threats

Already, this book has covered how cloud-based attacks are making businesses more vulnerable, especially as reliance on technology permeates more and more of modern society. Artificial intelligence, known colloquially as AI, is software that automatically makes important business decisions for you by using real-time, real-world data. From predicting customer behavior to self-starting fraud detection software, AI is important to modern businesses that want to stay one step ahead of the market.

Organizations that don't use AI for their threat exposure traditionally rely on signature-based detection. Essentially, they scan and detect some kind of unique attribute to identify what kind of threat it poses. This can be anything from a particular patch of code to a file attached to the malware. Signature-based detection is estimated to uncover 90% of cybersecurity risks. In 2019, the Institute of Electrical and Electronics Engineers found that upgrading to AI-based threat detection will bring up that number to 95%.

As AI gets better and better at predicting what's best for your company at any given time, this unfortunately gives cybercriminals room to use the same features against you. With AI technology, hackers can create malware that automatically attacks a system by finding its weak spots without human intervention. It's similar to the AI technology that scans and detects bugs in software, but turned against businesses to give hackers the upper hand.

It's not just malware detecting vulnerable entry points in your network at stake here. Once inside your system, AI enhanced threats can detect which would be the most lucrative attack to mount against your particular systems. Just by reading what's visible on your network, AI can fully learn the scope of your organization to launch malware made to hit your computer systems where it will hurt the most.

Since machines learn fast, your business needs to be prepared before anything happens. Make certain your team is up on the latest AI-focused cyberattacks so they know how to spot and avoid common pitfalls.

Reading this book is terrific start!

ANATOMY OF AN AI ENHANCED ATTACK

In a world of ever-evolving technology, the smarter your business's machines get, the more opportunities hackers have to steal information too.

For example, they can scan for vulnerabilities in the network so hackers can go straight for the backdoor and infiltrate your system. This is called **AI fuzzing**, and it's a useful tool for hackers now. They become robotic experts on the machine they've infected so they can quickly assess if it will be a useful entry point into your network. This significantly reduces the time it takes to break through your security systems which in turn gives you less time to launch an adequate and appropriate response.

AI enhanced cyberattacks are a good choice for hackers. They work automatically and can adapt quickly depending on what's found inside the victim's system. This ensures the malware causes maximum possible damage. Cybercriminals can hit fast, hard and where it hurts the most when they turn to artificial intelligence.

This isn't the only way that cybercriminals are getting into secured networks. AI enhanced attacks give them new opportunities that may surprise you, and you need to be prepared for anything.

Case Study: TaskRabbit
WHAT WENT WRONG?

Clients and customers are the most important thing, so their security should be your top priority. They rely on your discretion and abilities when they entrust you with their personal information. That's why data breaches can cause a massive hit to your reputation that's hard to get back once they feel unsafe spending time and money on you.

TaskRabbit learned this lesson the hard way. The website connects freelance laborers to clients who hire them to complete everyday tasks for an agreed upon price. In April 2018, 3.75M users found that their TaskRabbit profiles had been hacked. Victims had their social security numbers, bank account details, names and other personal information compromised in the attack, which hit both hirers and workers equally.

So what happened? TaskRabbit was affected by an AI fuzzing attack perpetrated by a botnet, which automatically scanned the servers for weak spots that they could exploit. Then, the botnet let in the hackers that released them in the first place. From there, they could freely steal the personal data from all of the users who had stored that information on their profiles.

What's more, when TaskRabbit shut down the site to handle the situation, 141M more users got their data stolen while the company's security sorted out the holes in the network that had let the hackers in. Eventually, they got the website servers back up and running safely.

For all the trouble caused, TaskRabbit offered affected users twelve months of free identity-restoration services as well as a year of free credit monitoring services to anyone whose social security numbers and bank accounts were put at risk during this cyberattack.

WHAT SHOULD HAVE HAPPENED?

Just as AI fuzzing was the downfall of TaskRabbit in this situation, you can use it in a similar fashion to your advantage instead. AI fuzzing can be used as a prophylactic to find vulnerabilities in your system before going live with updates or a new website, for example. Hackers are exploiting modern technology to perform deadly AI-enhanced strikes, so why not use the same technology to prevent them from hurting your business? Use AI fuzzing to scan your system for intruders and malware on a regular basis, rather than having it weaponized against you first.

Some businesses also use decoys or *"honeypots"* in the server to circumvent AI-enhanced threats; the cyberattacker's scan will lock onto one of these honeypots instead of actual data. This services two purposes: It prevents them from getting their hands on real, personal information and also notifies the business that there's been a breach so that you can take action to patch it up as soon as possible.

At least when TaskRabbit was compromised, they offered free services for their affected userbase that directly related to the damages they suffered. It's a good strategy for showing them your dedication to cybersecurity moving past this incident together.

When client information gets compromised, think of relevant ways to make it up to them. When AI-enhanced cyberattacks happen, businesses can preserve some of their reputation by being transparent and having an open, honest dedication to repairing the negative impacts caused by the attack.

Case Study: Instagram
WHAT WENT WRONG?

Instagram has been one of the most popular social media platforms since it launched ten years ago. The app is a way for people to share photos, videos and reels to their followers' "feeds." It's also a useful tool for businesses. Companies promote their services and products via stories, posts and reels; they can also chat directly with their customers through Instagram's direct messaging. In 2019, this popular platform experienced two massive cyberattacks that left its userbase reeling.

The first one occurred in August of that year. A group of Russian hackers took over hundreds of individual Instagram accounts by using AI fuzzing to scan for user data that would make those particular victims more vulnerable to exploitation. Armed with what they needed, the hackers broke in and changed the phone number and email address that was associated with each of the accounts so the real owners couldn't get back into their profiles. Then the website's users found themselves facing even further frustration, as it proved difficult to regain access to their accounts. They inundated Instagram's customer service department with support requests to regain access before the problem was solved.

Later, in the autumn, Instagram was hit by another cyberattack caused by a bug in the code that exposed users' passwords in the URL, giving hackers a red-carpet into their accounts.

It isn't known how serious the exposure was, but experts believe some 5 million users had their data stolen as a result. This is obviously a huge security flaw making 2019 perhaps the worst year for Instagram.

Take steps to prevent AI-enhanced threats from breaching your network by deploying them as prophylactic countermeasures so this doesn't have to happen to your business.

WHAT SHOULD HAVE HAPPENED?

To be safe in a time when hackers can scan a userbase filled with millions of accounts to find their individual, private information and steal their personal data, you need to learn how to protect your business against all manner of AI-enhanced cybersecurity threats.

First and foremost, improve or upgrade your security detection software to catch AI threats in the act and shut them down before they do real damage to your system. The faster you spot holes in your attack surface, the quicker you can respond. Since these hacks often change to fit their victim, their malware, traps and other forms of manipulation will be as enticing to their particular targets as possible. That makes it much harder for you to identify the threat before it strikes, which puts you at a particular risk compared to some other social engineering threats.

Learn what kinds of AI technology cybercriminals tend to rely on so you can prepare a proper defense against them more effectively. Remember to do intruder detection checks regularly on your system as well, and employ AI in your cybersecurity arsenal so you can catch criminals before they cause real damage to your system, your business and your userbase.

The more AI technology adapts to help businesses like you thrive, the more cybercriminals will learn how to bend it to their own will. Staying cautious and up-to-date on the latest trends in security will protect your business against severe damages like the ones exploited here.

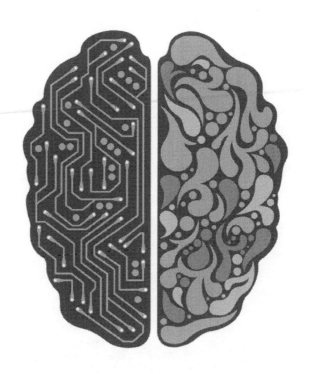

Chapter THIRTEEN

How Cybersecurity is Changing to Help

Earlier in this book, I warned you not to let your concerns about the prevalence of social engineering (and reverse social engineering) attacks get you down, and that's still true. Because as swiftly as hackers and attackers are coming up with ways to exploit smart contracts and scam you with new spoofs, experts in cybersecurity are working twice as hard to shore up your defenses.

Advancements in technology, after all, work both ways. As criminals are trying to break in, people are being paid a lot more money to stop them from doing just that. We're constantly coming up with new defenses against breaches, as well as offensive software capable of going after threats once they've been flagged on your system.

Here are some up-and-coming software, devices and capabilities that are rising to prevalence in the technology circles today.

ADVANCEMENTS IN CYBERSECURITY

Truth be told, AI defense mechanisms are much more ubiquitous than the social engineering tactics you fear. Although they use similar methods, and you should still remain wary of AI-enhanced threats, this technology is much more likely to be leveraged in your favor than against it.

You have a wealth of options to choose from when it comes to antivirus programs equipped with the kind of capabilities you're looking for, like AI fuzzing and protection from rogueware. It's much more likely that AI will protect you than harm you.

Other advancements to keep an eye on include:

- Virtual Dispersive Networking, or VDN, which is used to combat man-in-the-middle and eavesdropping attacks. Drawing in part from military technology, VDNs divide communications into a bunch of random pieces as it's passing over the network, which makes it difficult to read until it's reassembled at the proper endpoint.
- There is cloud monitoring software that detects unauthorized activity on a network during unusual times, like outside your hours of operation, or from suspicious personnel like a low-level employee opening a high-security file. This will alert your designated internal I.T. point person to investigate the suspicious activity, allowing you to quickly react after a breach.
- User behavior analytics, or UBA, monitors your employees' activities on a daily basis so they can alert you to unusual activity on their account. Much like the cloud monitoring mentioned above, this helps catch criminals who have compromised your employees' computers and taken advantage of their credentials. Even if they get past your firewall, they won't be safe from this.

New trends are growing alongside advancements in technology and security software. For example, it's becoming more commonplace for people to use complicated passwords and for websites to require a mix of letters, numbers and special characters in their passwords to make them harder to hack. Though that's a simple instance, it shows how, over time, cybersecurity becomes just as normal as the IoT devices we carry in our pockets everywhere we go.

Amongst antivirus software developers, the rise in more types of malware have increased their range of capabilities. The more threats they have to fight, the better fortified your environment becomes. We're learning more every day about how to protect our devices, so we're always on upwards path toward greater security.

Two- and three-step authentication factors are also becoming routine. Plenty of websites now require you to fill out a captcha, answer a security question or reply with a text code before letting you onto the site. This is in addition to mandating more complex passwords. All of these layers of protection, also known as defense-in-depth or DiD, reduce the risk of someone getting through to the network and systems and accessing private information simply by stealing your passwords.

As the workforce shifted en masse to their homes during the pandemic, companies had to contend for the first time with a staff that was working partially or even completely remote. This presents obvious cybersecurity risks, given that your staff's personal devices and home network may not be as secure as you would like, or may be shared with other family members who don't know what they should protect or what you consider confidential or private. With these obstacles unavoidable, it's become more commonplace for businesses to provide extra protections for their employees' IoT devices, even, in some instances performing stress tests on them.

Behavior like this increases cybersecurity both on-premises and off. It's all part of a culture of security that businesses should, no *must* implement if they want to stay protected from the latest cybersecurity threats.

KEEPING YOUR BUSINESS SECURE

Now you know everything there is to know about social engineering threats, reverse social engineering threats and how to stay safe from both with the most up-to-date cybersecurity measures available today. Protecting your business from the twelve major cyberattacks that we outlined in this book by staying informed about the latest news on cyberattacks, threat prevention and everything else going on in the industry today is a terrific first step in creating a culture of security in your business; a culture that keeps awareness top of mind and vigilance becomes second nature. You never know when something like what we've detail here could happen to your business.

To recap, here is a list of tips for keeping your business safe from the biggest socially engineered cybersecurity threats prevalent in the industry today:

- Learn and stay current on the common deception techniques that cyberattackers use to try and gain access to your system, such as spying on social media and relying on common courtesy to get them deep into your organization.
- Don't let fear or temptation motivate you into making mistakes that you wouldn't with a clearer head.
- Be careful where and to whom you entrust identifying information and bank account details, even if you've done business with these parties before.
- Don't let convenience cause you to drop your guard. Never store password in your browser or turn on

auto-fill as it may store important information like your credit card or passcodes.
- Trust but verify; never hand over money unless you're positive who's on the other end of the transaction, add a phone call verification to any monetary transaction.
- Store data back-ups in different places. As the saying goes, it's best not to put all your eggs in one basket.
- Invest in employee training for every level of the organization, so all employees are equipped with the cybersecurity knowledge necessary to keep your business safe.
- Use AI fuzzing to discover vulnerabilities in your network before botnets do the same in an attempt to exploit you. Combine AI security with more traditional, signature-based detection techniques, to catch as many bugs and holes in the system as possible.
- Get the latest in cybersecurity defense mechanisms so you're ready against evolving social engineering threats as they come.

Most importantly, nothing beats plain old common sense. Your boss shouldn't ask you to email over bank account details that they already have on file. You probably didn't win a free iPad by being the millionth visitor to a site. And if your computer is running just fine, there's no need to listen to that suspicious phone call insisting that it needs a remote security update with no time to waste.

When it comes to sending money, revealing personal information, relaying important business-related files, and cybersecurity in general, ask yourself this simple question: Does this make sense? Or would I be better safe than sorry?

This Photo by Unknown Author is licensed under CC BY-NC

Chapter FOURTEEN

Technical Terms Explained in Plain English

ARP spoofing – when cybercriminals intercept address resolution protocol requests and send false data in their place, ultimately linking their MAC address to a legitimate IP on the server

Baiting attack – a ruse where cybercriminals promise a tempting prize or make claims that pique the victim's interest

Cloud computing – mass storage and processing of data and information in a remote server that the user can access anywhere and anytime

DDOS attack – distributed denial-of-service attack, or DDOS attack, happens when cybercriminals flood website servers with requests to disrupt traffic until they stop working altogether

DNS spoofing – cybercriminals falsify a domain name to catch and redirect traffic to a secondary, malicious website with fake credentials

Domain spoofing – when cybercriminals impersonate a particular email domain by using a one with hard-to-detect differences to fool employees

Eavesdropping – common type of man-in-the-middle attack where criminals secretly monitor communications from real parties to glean confidential information from them

Firewall – a block in your computer system that prevents malware from getting through and users from gaining access to certain information and websites

IP spoofing – when cybercriminals falsify their IP address to send network packets to other computers

Man-in-the-middle attack – a cyberattack where real messages are intercepted and potentially changed before the recipient gets it

Modifying traffic – common type of man-in-the-middle attack where criminals secretly intercept and alter genuine messages from parties within a network

Phishing – a type of hacking where the cyber-criminal steals private information by tricking a you or somebody else into giving it up, usually through spam or some type of theft of your account information

Pretexting – when a cyberattacker fosters a false sense of trust with the victim, usually by pretending to be some kind of authority figure to convince you to give over your social security number or other personal information

Secondary attack – the cybercrime that follows up an initial breach or entry into your computer network

Social engineering attack – cybersecurity threats that result from an untrustworthy source tricking someone in the network into giving out compromising information

Spoofing attack – a cyberattack where criminals falsify their information to seem legitimate somehow, so their intended victims trust the source of the data

Tailgaiting – a cybersecurity threat where the attacker follows authorized personnel into secured sectors to gain access to private data they wouldn't otherwise be able to reach

Remote access trojan (RAT) – malware that downloads secretly on a computer to open a backdoor, for the hacker to remotely take control over the machine

Reverse social engineering attack – cybersecurity threats that result from someone in the network unintentionally establishing contact with an untrustworthy source and then giving out compromising information

Water-holing attack – a cyberattack where the criminal gathers information about what websites their targets spend the most time on and use those landing pages to infect the users' computer with malware

An Invitation to the Reader

The reason I published this book was to fortify businesses and educate business owners with the basic knowledge they need to make a great decision when choosing the right MSP or CISO. I believe a qualified MSP can contribute greatly to your business success just like a great marketing company, attorney, accountant or financial advisor.

Cybersecurity as a specialty is so new, and growing at such a rapid pace, that most business owners can't keep up with all the latest whiz-bang security gadgets, alphabet soup acronyms, and choices of protection available to them. Plus, with some 80,000 new strains of attack being released daily, many of the "latest and greatest" technological developments have a shelf life of a few days, weeks or months before they become obsolete or completely out-of-date. Sorting through this rapidly-moving mess of information and ever-changing threat landscape to formulate an intelligent cybersecurity plan for to protect and grow a business requires a cybersecurity professional who not only understands the treat landscape and the technology to harden the attack surface, but also understands how people and businesses need to interact with this technology and allow them to be protective while being safe.

Unfortunately, the complexity of technology, the special skillset required to understand the threat makes it easy for a business owner to fall victim to a cyber-attack. Incompetent or dishonest security companies or consultants, or well-meaning IT people who aren't specialists in cybersecurity create gaping holes in a company's security posture. When this happens, it is not a question of if, but a question of when you will become the next victim.

Therefore, my purpose is to not only give you the information you need to find an honest, competent technology company or consultant with the proper cybersecurity skills to support and protect you, but to enlist you and your team as soldiers in the fight against cybercriminals. When you know what to look for and how to protect yourself, you become part of the protective fabric that stops cybercriminals and hackers by preventing them from practicing their craft. I believe that, the more this topic is discussed and the more educated we all are, the safer it will become for everyone.

I certainly want your feedback on the ideas and information in this book. If you try the strategies I've outlined and they work, please send me your story. If you've had a bad experience with a cybersecurity company or consultant, I want to hear those horror stories as well. If you have additional tips and insights that we haven't considered, please share them with me. I might even use them in a future book!

Again, the more aware you are of what it takes to protect your business and clients from cybercrime the stronger your business will become. I am truly passionate about building a secure organization that delivers uncommon service and security to my customers. I want to help business owners see the true competitive advantages secure technology can deliver to your business, and not just view it as an expensive necessity, compliance requirement and source of problems.

Your contributions, thoughts and stories will make it possible. Please write, call or e-mail me with your ideas.

VALUABLE COUPON FOR DISCOUNTED BUSINESS SECURITY RISK ASSESSMENT

(Minimum $1247 Value)

Don't let your team or yourself become another casualty of a cyber-attack! If you are a business owner who believes that securing your, your business or your customers data is important, I'd like to offer you 25% discount on a Business Security Risk Assessment that will:

- Determine if your employees are putting you at risk.
- Scan your public assets and social media accounts for at-risk profiles.
- Audit existing policies and procedures for gaps or inadequacies.
- Perform vulnerability assessment on internal assets.
- Review in-practice outcomes for alignment.
- Determine if patterning can be established.
- Evaluate layered security measures for use and enforcement.

25% OFF

BUSINESS SECURITY RISK ASSESSMENT COUPON (Minimum $1297 Value)

Call our Security Team at 855-255-1550 and just say that you want the FREE Acceptable Use policy and 25% OFF a Security Risk Assessment.

No duplicate coupons or copies will be accepted.
This offer can't be sold, traded or combined with any other offers.

BOOK ORDER FORM

If you enjoyed this book, share it with others! Use this form to order extra copies for friends, colleagues, clients or members of your association. Please allow 2-4 weeks for delivery.

Quantity Discounts:
1-9 copies = $14.95, 10-49 copies = $11.95, 50-99 copies = $9.95
100 or more copies = Call for discounts and wholesale prices

Information:
Name: _____

Company: _____

Address: _____

City: _____

State/Province: _____

ZIP/Postal Code _____

# of copies _____ @ $_____ Total: $_____

Add shipping and handling @ $3 per book: $_____

Please make check or money order payable to Absolute Logic and mail to:

        Absolute Logic
        88 Danbury Road, Suite #1D
        Wilton, CT 06897

Credit Card: ❑ Visa ❑ MC ❑ Amex ❑ Discover
Card Number: _____ CVV: _____
Expiration Date: _____
Signature: _____

Thank you for your order!

Made in the USA
Middletown, DE
23 June 2023